MODELLING RADIOCESIUM IN LAKES AND COASTAL AREAS – NEW APPROACHES FOR ECOSYSTEM MODELLERS

Modelling radiocesium in lakes and coastal areas – new approaches for ecosystem modellers

A textbook with Internet support

by

Lars Håkanson
Uppsala University,
Sweden

SPRINGER-SCIENCE+BUSINESS MEDIA, B.V.

A C.I.P. Catalogue record for this book is available from the Library of Congress.

ISBN 978-94-017-3857-6 ISBN 978-0-306-46878-0 (eBook)
DOI 10.1007/978-0-306-46878-0

Printed on acid-free paper

CONTENTS AND PROLOGUE

The lake model for radiocesium is based on work carried out in the ECOPRAQ-project (an EU-project co-ordinated by Rob Comans). The coastal model has been done in the MOIRA-project (EU-project co-ordinated by Luigi Monte). I would like to express my gratitude to the colleagues in those very interesting projects. Without their support and constructive input, this book would not have been published.

CONTENTS AND PROLOGUE

CONTENTS

CONTENTS AND PROLOGUE

PROLOGUE

There are some specific objectives for this book. First, the idea is to present a state-of-the-art model for radiocesium in lakes. This model has been validated against independent data for lakes covering a wide domain of lake characteristics, and it has yielded very high predictive power for the target variables, Cs-concentrations in water and fish. Practical usefulness in contexts of lake management and research involves at least three things: high predictive power (otherwise the simulations are mere theoretical exercises), easy access to the necessary driving variables, and a wide applicability. This model probably meets these three requirements better than any other model for water pollutants. The predictive power of any given model is determined not by "logical and convincing" arguments but by equations and coefficients (Peters, 1991). A crucial element in the derivation of the radiocesium model for lakes has been to find the most relevant general structure and the best possible equations related to this structure handling the most important fluxes of radiocesium to, from and within lake systems. This model also meets two requirements that may seem contradictory: few and readily available driving variables, and a relevant general ecosystem structuring to handle all important radiocesium fluxes. Solving this paradox is the key to the predictive success of this lake model. The aim of the first section of this book is to present a solution to this seemingly paradoxical task. This model handles the following abiotic processes: fallout on the lake surface and catchment area, transport from catchment area to lake, transport in dissolved and particulate phases, sedimentation, internal loading (advection and diffusion), transport to passive (geological) sediments, bioturbation (mechanical mixing) and outflow. The following biotic processes are covered: biouptake, excretion, biomagnification, feeding habits, temperature dependencies, fish weight dependencies, biological dilution, and radiocesium uptake by cells and competition with potassium. In spite of the fact that all these processes and mechanisms are accounted for, the model is driven by just a few readily available variables, like latitude, altitude, catchment area, characteristic soil type, lake area, mean depth and K-concentration. Such data are generally available for lakes from standard monitoring programmes. Many of the structures, sub-models and equations are general in the sense that they apply to other water pollutants than radiocesium and to other aquatic systems than lakes.

The key components, equations and motivations for the lake model have been presented before (Håkanson, 1999). However, there are some pertinent changes between the two models concerning the equations for internal loading, symbols and abbreviations and, above all, motivations and critical tests.

The second part of this book concerns a model for radiocesium in coastal areas. This model is based on several sub-models used for the lake model. However, the main problem in applying these sub-models for lakes also to coastal areas concerns the very definition of a coastal area. The lake model is based on ordinary differential equations (i.e., it is a compartment model) using a calculation time (dt) of one month to obtain seasonal variability. Basically, it is a dynamic mass-balance model. It is easy to use such models for entire lakes, which are well-defined geographical units with clear borders. But how is it possible to use compartment models for open coastal areas? Section two presents a method for defining the borderlines of coastal areas (the topographical bottle-neck technique). This means that morphometrical parameters like mean depth, relative depth, etc. can be defined and used in predictive models in a meaningful manner. It also means that parameters of fundamental importance in mass-balance modelling to calculate concentrations, fluxes and amounts, like water retention times for surface water and deep water (to handle water transports) and bottom dynamic conditions (to handle internal loading), can be defined for the given coastal areas. These fundamental modelling prerequisites can then be predicted from models based on morphometrical parameters. The morphometrical parameters can, in turn, be easily determined from bathymetric maps using GIS-methods (GIS = Geographical Information Systems).

A very important part of this book has been to illustrate how methods for critical model testing, like sensitivity analyses and uncertainty analyses using Monte Carlo simulations, can be important tools not just in the traditional testing of models but also in the building and simplification of models. Aquatic systems are extremely complex webs of physical, chemical and biological interactions; "everything depends on everything else". Every process can be mathematically described in different ways in such complex ecosystems and empirical data are absolutely essential for calibrating and validating models. But empirical data are always more or less uncertain because of deficiencies in sampling, transport and analyses. They may poorly represent the sampling bottle from which they emanate and the sampling bottle may poorly represent the coastal area it is supposed to represent. The accuracy and uncertainty of predicting target y-variables depend on such structural and empirical uncertainties. But all uncertainties cannot be of equal importance in predicting the target y-variables. The methods discussed in this book are meant to be useful for quantitative ranking of such uncertainties so that models can be built on the most important and relevant model variables, no more - no less.

Another novel idea with this book concerns the Internet support. The reader of the book can go to the following internet address: http://go.to/csmod. There one will find information concerning the software, "I think", "Stella" and "PowerSim" as well as the lake model, the coastal model and databases. This means that the user will be able to follow the model structuring, and the given equations as she/he reads the text, and also test the models for other lakes and coastal areas and make simulations about remedial measures or other applications. Hopefully, the possibility of Internet support will create a forum of contacts between the users of these models and similar models, so that better data will be available for, e.g., calibrations and validations of the coastal model and further improvements of our modelling skills. This would be important, and we could probably all agree on this, not just for water management but also for science since modelling basically concerns our understanding of ecosystem structures and functioning.

CHAPTER 1

MODELLING RADIOCESIUM IN LAKES

1.1. Introduction and Aims

1.1.2. BACKGROUND ON RADIOCESIUM AND RADIATION DOSE TO MAN AND ENVIRONMENT

The nuclear accident at Chernobyl resulted in large parts of Europe being exposed to a major fallout of radionuclides in late April-early May, 1986 (Persson et al., 1987; Moberg, 1991).

"....if man is adequately protected then other living things are also likely to be sufficiently protected". This statement from the International Commission on Radiological Protection (ICRP, 1977) is based on extensive reviews on the effects of ionising radiation, including radiocesium (USNRC, 1978; IAEA, 1992; ICME, 1995). Today there are many ongoing studies within the framework of international organisations (such as IAEA, the International Atomic Energy Agency, Vienna) and several EU-projects (such as MOIRA, see Appelgren et al., 1996; Monte et al., 1997), to structure the existing knowledge and do research in the field concerning radiological effects to ecosystems.

The effects of ionising radiation on aquatic organisms concern both individuals and population levels and the variables of interest are, e.g., mortality rate, fertility rate and mutation rate (see Whicker and Schultz, 1982; Jimenez and Gallego, 1998). There are major differences between different organisms in terms of sensitivity to ionising radiation, and there are major differences for one type of organism living in different environments.

Ionising radiation can result in damage at cellular and whole-body level. Molecular changes may be induced in the DNA. When effects such as chromosomal alterations do not result in cell death, they can be transmitted to subsequent generations. The response to the radiation can be modified by several factors, such as the type of radiation (different types of radiation, such as α, β, γ, have different biological effectiveness) and the dose rate (a higher total dose is required at low rates to produce the same effect than at high dose rates). The existence of a threshold of dose rates above which effects are produced has been the subject of a long debate. DNA repair, which operates at a molecular level, may have a particular significance at low dose rates.

Environmental factors can also influence responses to radiation, e.g., in mammals. Radioresistance increases under anoxic conditions. Salinity tends to increase radiosensitivity because of the metabolic stresses they impose on the organism. Radiation response also depends on the stage of the cycle the cell is in when the exposure occurs. In general, the radiosensitivity of cells is directly proportional to their rate of division and inversely proportional to their degree of differentiation (IAEA, 1988).

There are also indirect effects of radiation, e.g., more resistant organisms can gain a competitive advantage over other less resistant groups. The difficulties of getting meaningful results about radiation effects on natural populations in their natural environments have limited the available information on many species and communities, and very little information has been obtained about the effects of interactions between radiation and other agents or stresses, and the long-term effects of low-level radiation.

Aquatic organisms will receive external exposure from radionuclides in water and sediments and internal exposure from nuclides accumulated in their tissues. The calculation of such doses is the goal of several dosimetric models (IAEA, 1988), which use as input the amount of radionuclides in the water.

In the absence of well-tested target effect variables related to the reproduction or survival of key functional organisms in real aquatic ecosystems, the concentration of radiocesium in fish muscle consumed by man is generally used as a simple operational effect variable, and the guideline value for commercial marketing of fish recommended by the National Food Administration (e.g., 1500 Bq/kg ww in Sweden) is often used as a reference value and in setting environmental goals.

1.1.3. BACKGROUND TO THE MODEL FOR RADIOCESIUM IN LAKES

Nutrients, metals, radionuclides and many toxic substances are distributed in aquatic ecosystems by the same fundamental processes. This means that the same principles of modelling can be used for many chemicals, pollutants and/or natural substances in most aquatic systems, and this modelling structure can be increased almost indefinitely by incorporating more processes for fluxes (mass/time), amounts (mass) and concentrations (mass/volume) to and from the defined compartments. There are many such processes. The aim here is NOT to give a thorough account of all these processes or to discuss the many methods and criteria that may be used to critically test models, since there are many textbooks already available on that subject (see, e.g., Vemuri, 1978; Jørgensen et al., 1986; Jørgensen and Johnsen, 1989; Straskraba and Gnauck, 1985; Schulze and Zwölfer, 1987; Håkanson and Peters, 1995).

The dynamic model presented here for radionuclides in lakes focuses on basic principles and processes and accounts for internal processes (diffusion and advection), seasonal variations on a monthly basis, and dissolved and particulate fractions of radiocesium; it also relates the calculated lake concentration of radiocesium to concentrations in biota (here fish).

Predictive models are fundamental tools for forecasting the consequences of different remedial actions (see IAEA, 1999). The term "predictive model" is used to indicate models in which one or a few important y-variables are predicted from a few x-variables that can be obtained easily from standard maps or monitoring programmes. That is the goal, but that goal is often very difficult to achieve. It is as much a challenge to develop practical models as it is to create comprehensive validated ecosystem models that predict the time-dependent interactions of many y-variables from many x-variables.

There are interesting general principles that can be addressed using radiocesium as type substance, namely the transport related to a sudden peak in contamination, the following fundamental differences between pelagic and benthic transport routes, the importance of the partition (= distribution) coefficient and the retention times in different compartments.

Many of the processes regulating the flow and biological uptake of substances in lakes are controlled by hydrological and morphometrical factors, which cannot be influenced by remedies changing the water chemistry, e.g., liming and potash treatment (Andersson et al., 1991; Håkanson and Andersson, 1992; Abrahamsson and Håkanson,

1997). Remedies that could speed up the recovery (e.g., reduction of the elements in fish) must aim at, e.g., reducing the load to the lake, or the uptake in biota by blocking the biouptake, increasing bioproduction and cause a "biological dilution" of the elements in biota, and/or include specific substances, like potash (K), that can be taken up in fish in almost the same way as radiocesium and cause a "chemical" dilution. Time is an important factor in such remediations. The sooner the remedy is launched the better the results generally gained (Andersson, 1989). Lakes suited to potash treatment should have low initial ionic K-concentrations (Dahlgaard, 1994).

How is radiocesium transported in soil, water and biological material? Which lakes are particularly sensitive to radiocesium contamination? Which measures could speed up the recovery? Such questions can be answered by the following model.

1.2. The Modelling Principles for the New Model for Radiocesium in Lakes

Practically useful models must satisfy some categorical features that make them reliable tools for environmental management:

- they must be characterised by a relevant and simple structure, i.e., involve the smallest possible number of driving variables;
- the values of the necessary driving variables should be easy to access and/or to measure;
- the models must be validated for a variety of circumstances showing a wide range of environmental characteristics.

In broad terms, the variables used in environmental models may be divided into two categories:

1) variables for which site-specific data must be available, such as lake volume, mean depth, water discharge, amount of suspended particulate matter in water, etc.;

2) variables for which generic (= general) values are used because of the lack of site-specific data, e.g., the sedimentation rate and/or rates for internal loading (like diffusion and advection rates).

The variables belonging to the first category are often called "site-specific variables", "environmental variables" or "lake-specific variables". They can generally be measured relatively easily and their experimental uncertainty should not significantly affect the overall uncertainty of the model predictions of the target variable(s). The second category, the "model variables" or "model constants", are often difficult to access for each specific system, such as the transfer rates from the sediment to the water, the deposition velocity of X from water to sediments, the migration rate from catchment to lake, etc. The model variables may contribute significantly to the model uncertainty unless they have been validated and reliable sub-models for their generic values established from critical tests.

Unfortunately, critically validated widely applicable models showing a high predictive power are difficult to develop. As a consequence, practically useful models are generally (see Monte et al., 1997) based on "collective parameters". It has been shown that (Monte, 1996) in many circumstances the values of such important driving collective parameters integrate many compensation effects of the different phenomena occurring in the very complex ecosystem where "everything depends on everything else" (Håkanson and Peters, 1995). Examples of such collective parameters in the freshwater environment are the "migration velocities" of the metal or radionuclide from water to sediments, the "effective removal" rates and the "soil permeability coefficient" of a radionuclide from the catchment to a water body (Monte, 1995).

Models based on such "collective parameters" show a unique and important feature: their predictions are characterised by a relatively low uncertainty, despite the large range of environmental characteristics and the lack of site-specific values of the model variables (see Håkanson, 1995). The use of models based on collective parameters is the result of a change in the perspective concerning the modelling of natural ecosystems (see Håkanson and Peters, 1995). The main lesson is that in predictive modelling, it is seldom necessary, or wise, to account for "everything". The difficult task is to omit processes which may add more to the model's uncertainty than to the predictive success for the given target variable.

Environmental models based on collective parameters are structured more like a web than a pyramid. Each process is indeed related to a variety of other phenomena and it would be unreasonable to use only a few of them as fundamental starting points for understanding and predicting all the others. A variety of past experiences (see Peters, 1991; Håkanson and Peters, 1995) demonstrate that big complex models are often more uncertain, and yield a poorer predictive success than simpler models based on collective parameters.

It is often extremely difficult to distinguish cause and effect in natural ecosystems. One cannot base a dynamic ecosystem model on a full understanding of the ecosystem. In complex ecosystems "understanding" at one scale (e.g., the ecosystem scale) is generally related to processes and mechanisms at the next lower scale (e.g., the scale of individual animals and/or plants), and the explanation of phenomena at this scale is related to processes and mechanisms at the next lower scale (e.g., the scale of the organ), and so on down to the level of the atom and beyond. In environmental management, the predictive ecologist must often find a balance between answering interesting, often important, questions of understanding, and delivering a practical tool to society. If a model were based on a causal analysis of what takes place at the cellular level, then at levels involving organs, individuals, populations, and finally at the ecosystem level, one would wait an eternity before the model could be developed.

In a strict sense, there is no such thing as a general (= generic) ecosystem model that works equally well for all ecosystems (of a given type) because all models need to be tested against reliable, independent empirical data and the data used in such validations must of necessity belong to a given restricted domain. If this domain is equal to the entire population of ecosystems of the given type, then, and only then, is the model generic in the strict sense. The complexities of natural ecosystem always exceed the complexity and size of any model. Simplifications are always needed, and this entails problems. The dynamic mass-balance model discussed here has been tested over such wide ranges that it is tempting to label it generic, but there will always be an ecosystem with properties outside the given domain for which the model would yield poor predictions. This is why modelling can be pictured as a two-sided coin: one needs the equations as well as the range within which the equations apply.

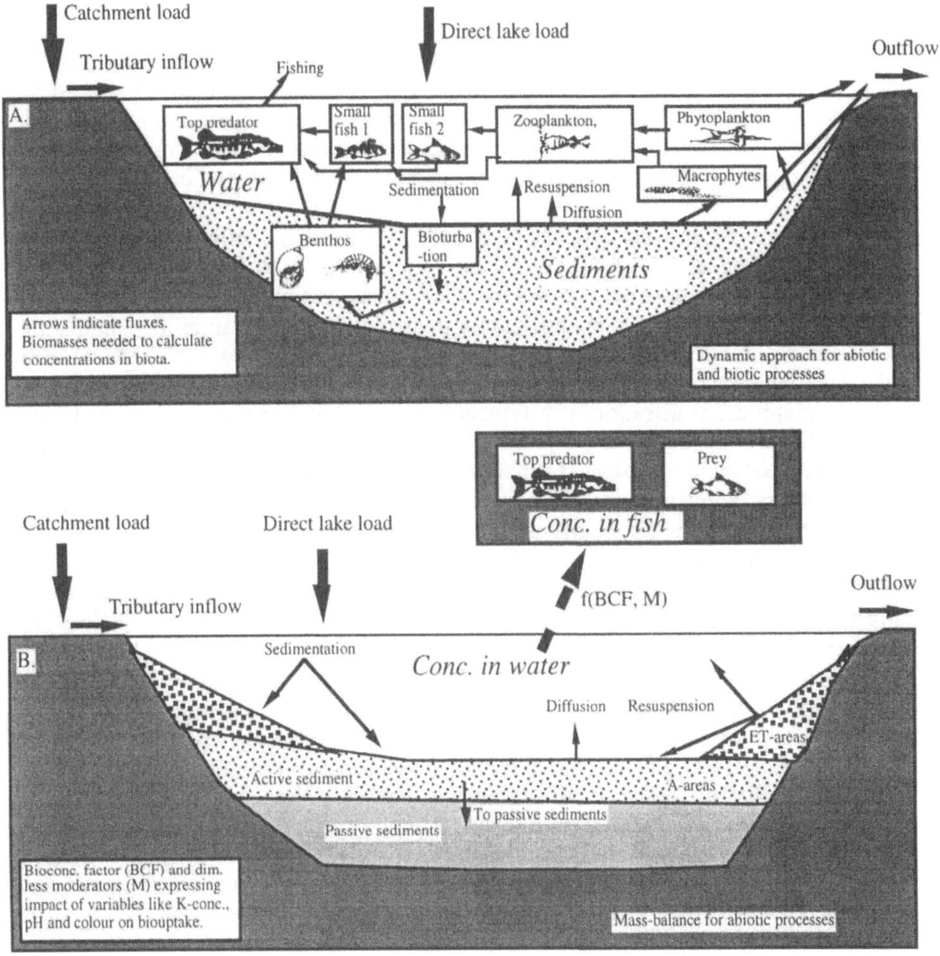

Fig. 1.1A. Compartment model illustrating the fluxes (arrows; mass per unit time) in a traditional dynamic model for a given metal in a lake ecosystem with compartments for top predator, two types of small fish, zooplankton, phytoplankton, macrophytes, benthos, water and sediments.
B. Illustration of a "mixed" model, i.e., a model based on a mass-balance concepts for all abiotic fluxes, a bioconcentration factor which is used to calculate concentrations of toxic substances in biota (so that one does not need to determine the biomasses, which is very difficult), and dimensionless moderators used to quantify how different water chemical conditions (pH, colour, K-concentration, etc.) affect biouptake. Modified from Håkanson (1999).

The requirements set out for this model are that it should:

- be mechanistically sound, i.e., whenever possible be based on fundamental scientific principles;
- include key processes regulating the transport of radiocesium, and other pollutants as well, in lake ecosystems;

• be possible to validate all essential parts or sub-models against practically accessible empirical data from the range of lakes for which the model should be applicable;
• have a high predictive power. This is generally controlled not by the strength of the best sub-models but by the weakness of the structural characteristics of the overall model and/or the strength of the weakest sub-models. This means that, ideally, all parts or sub-models of the overall lake model should be equally strong (or equally uncertain). That is, the model should be well balanced.

The new model discussed here is based on elements from many other lake models for radionuclides and heavy metals (see IAEA, 1999; Håkanson, 1999). The great challenge is to construct a model based on causal principles and an understanding of key processes, and then to simplify those parts that need to be simplified so that the model can be practically useful. This means that there is a logical development from understanding to simplification.

A graphical illustration of a typical compartmental model is given in Figure 1.1A. The figure illustrates the top predator (like pike or large perch), small fish 1 and 2 (e.g., small perch and roach), zooplankton, phytoplankton, macrophytes and the benthos, abiotic compartments (like water and active and passive sediments) and the processes regulating the fluxes between these compartments. The figure also gives the fluxes to the lake (direct lake load and river input related to catchment load) and from the lake (outflow and sediment transport from the active to the passive sediment layer). This is a general modelling set-up applicable to any substance X in a lake. The concentration of X in any biological compartment is defined as the ratio between the amount of X in the compartment (in g or Bq) divided by the biomass of the species (e.g., kg pike in the lake). One needs biouptake rates and excretion rates for all biological compartments, variables to describe the compartments (like age, sex, organ or weight of the fish) and the biomasses of the species in the lake. All these model variables are difficult and expensive to access, they vary seasonally and are often just guessed. This means that models of this kind generally provide poor predictions. Better predictions can often be obtained from models structured in such a way that one uses a dynamic model for all important abiotic processes, since such processes are relatively easy to handle compared to biological processes, so that the concentration of X in water and/or sediments can be predicted accurately. Then one can use the bioconcentration factor (BCF) approach to calculate the concentration of X in biota directly without having to estimate biomasses. Often there is knowledge about the factors influencing the biouptake, like a lower biouptake for radiocesium in lakes with high concentrations of potassium, or lower biouptake of mercury in lakes with high concentrations of selenium. This knowledge can be expressed by dimensionless moderators operating on the BCF-values (see Figure 1.1B).

The crucial point in practical lake-ecosystem modelling is the structuring of the model (Figure 1.2). The model structure should include all important processes regulating the transport and biouptake of X - no more, no less. Figure 1.3 gives the basic structure of the dynamic model for substances in lakes, which will be elaborated in detail in the following text. There are some fundamental structural components for all models of the type given in Figure 1.3. When there is a partitioning of a flow from one compartment to two or more compartments, this is handled by a partition (= partitioning = distribution) coefficient. This could be a default value, a value derived from a simple equation or from an extensive sub-model. There are five such distribution coefficients (DCs) in the following model:

1. The DC to distribute the fallout on the catchment area to inflow and outflow areas.
2. The DC handling the partition from inflow areas to outflow areas and lake water.
3. The DC differentiating between dissolved and particulate fractions. This DC is often referred to as the lake Kd-value.
4. The DC regulating the sedimentation to areas of erosion and transport (ET-areas) and/or to areas of accumulation (A-areas).
5. The DC describing the flux from ET-areas back to lake water and/or to A-areas.

The following model for radiocesium has been constructed as a tool (1) for modelling the time-scales for the "peak and the tail" for radiocesium concentrations in lake water, sediments and biota, (2) for facilitating interpretations of remedial measures for high levels of ^{137}Cs in fish (see IAEA, 1999), and (3) for highlighting the most important processes regulating spread and biouptake of metals and radioisotopes in lake ecosystems more generally. There are 22 rates and 18 model variables in the model in Figure 1.1A. To make a good prediction, one would need reliable lake-specific empirical data on all 40 model variables. Rates are sometimes called rate constants, but that might give the wrong impression that rates do not vary in time and space. Most rates are variables, just like weight and age of animals. This means that it is a great challenge to develop predictive models for the most fundamental rates and model variables, like lake Kd, the rate of sedimentation, and biouptake and retention rates.

Very many studies have been done on radiocesium in lakes after the Chernobyl accident in April/May of 1986. It is evident that many factors and processes can influence the variability in radiocesium concentrations in water and biota within and among lakes. But what relative importance should one expect for the various factors and processes? It is evident that hundreds of x-variables may be of interest, but they cannot all be of equal importance for the predictions of the target y-variable. Only reliable empirical data can be used to provide a ranking of different x-variables influencing a defined target y-variable. Table 1.1 can be regarded as an introduction to the following modelling. It gives (based on data from 14 Swedish lakes; see Andersson et al., 1991; Håkanson, 1991; Håkanson and Andersson, 1992) an r-rank table (based on linear correlation coefficients of absolute values) for one of our target y-variables, the concentration of radiocesium in pike (Cs_{pi88}) in 1988 in relation to:

1. Fallout, Cs_{soil} in Bq/m^2.
2. One variable for the load of cesium to the lake, Cs_{wa87} (Cs-concentration in lake water in 1987 in Bq/l).
3. Different lake variables (mean annual values for 1987) for colour (mg Pt/l), K-concentration (in meq/l), hardness (CaMg in meq/l), pH and total-P-concentration (totP = TP in µg/l).

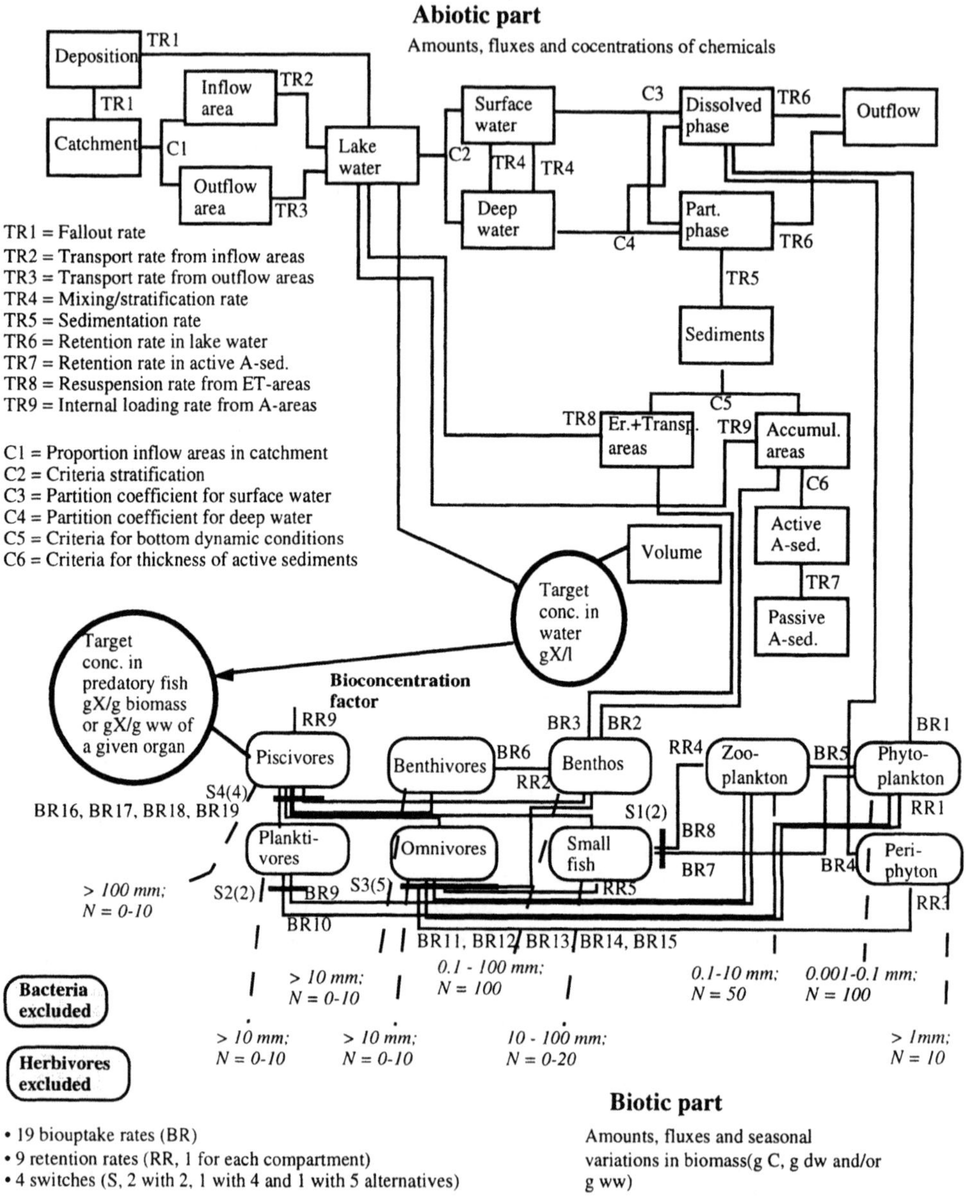

Fig. 1.2. A compilation of fundamental abiotic and biotic processes regulating the spread and biouptake of toxic substances in aquatic ecosystems. The figure also gives typical sizes (mm) and numbers of species (N) in Nordic lakes. This structuring of the aquatic ecosystem will be used in the following model, where the concepts will be further elaborated. Modified from Håkanson (1999).

Basic structure of the mixed model for radiocesium

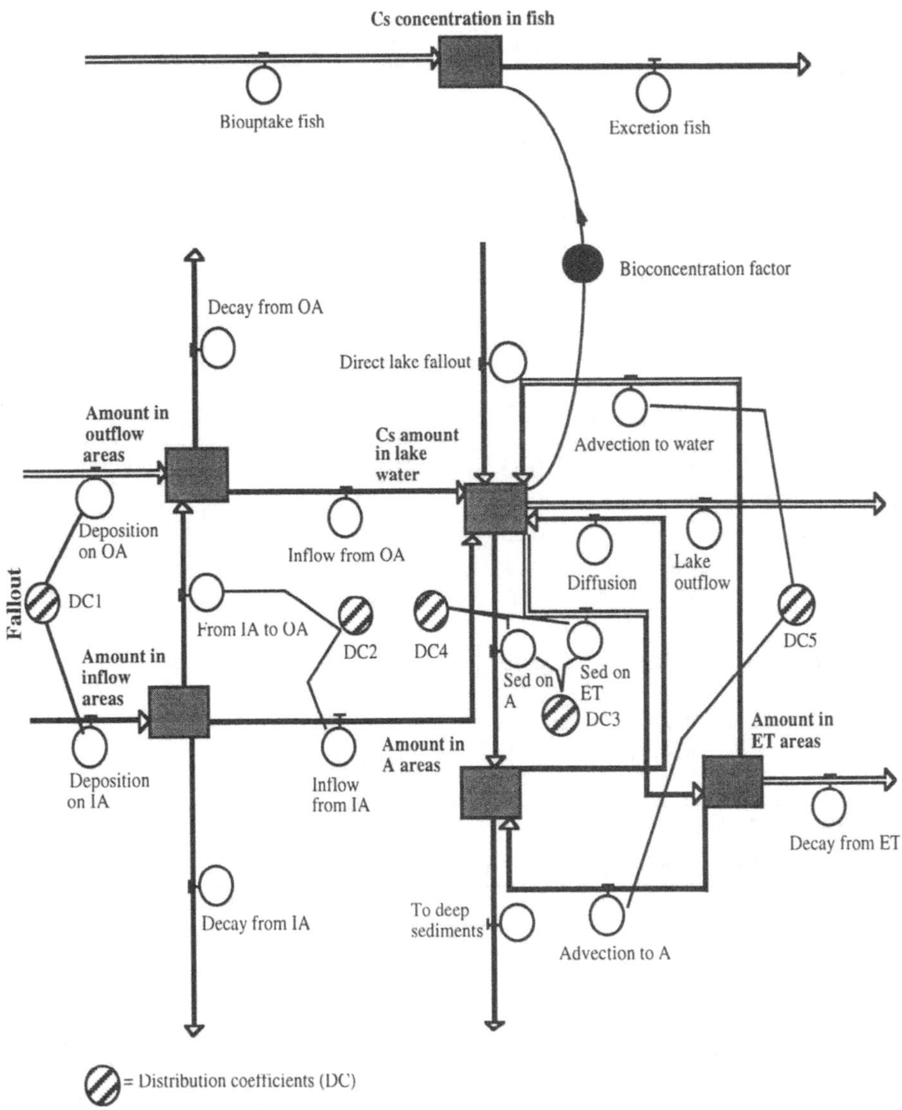

Fig. 1.3. Illustration of the basic structure of the mixed model for toxic substances in lakes.

Table 1.1. A. r-rank (linear correlation coefficient, r) matrix based on data from 14 Swedish lakes on cesium in pike in 1988 (Cs_{pi88} in Bq/kg ww) versus different variables.
B. "Ladder" from stepwise multiple regression (from Håkanson, 1997) with Cs_{pi88} as target y-variable.

A. r-rank;	**n = 14**		
		Cs_{pi88}	**CV**
y-variable		1.00	0.22
Fallout	Cs_{soil}	0.70	0.10
Lake load	Cs_{wa87}	0.88	0.26
Water variables	Colour	-0.20	0.19
	K	-0.31	0.12
	CaMg	-0.37	0.14
	pH	-0.42	0.02
	totP	-0.48	0.38
Morphometry	Dm	0.53	0.01
	Q	0.38	0.10
	Tw	0.23	0.10
	A-areas	-0.47	0.05
	DR	-0.64	0.02
Catchment	Rock%	0.40	0.01
	ADA	0.37	0.01
	RDA	0.12	0.01
	Open land%	-0.11	0.01

B. Ladder; $y=Cs_{pi88}$; n=14

Step	**F-value**	**Variable**	**r^2-value**	**Model**
1	4	Cs_{wa87}	0.78	$y=9479 \cdot x1+769$
2	4	K	0.89	$y=9559 \cdot x1-170.6 \cdot x2+2524$
3	2	Open land%	0.92	$y=9685 \cdot x1-249.5 \cdot x2+172 \cdot x3+2804$
4	1	totP	0.93	$y=9259 \cdot x1-226.4 \cdot x2+191.6 \cdot x3-224.6 \cdot x4+4939$

Cs_{pi88} in Bq/kg ww; Cs_{wa87} in Bq/l; K in µeq/l; Open land in % of ADA, area of drainage area; Dm = mean depth in m; Q = mean water discharge in m^3/s; Tw = theor. water retention time in yr; A-areas = accumulation areas; DR = dynamic ratio; RDA = relief of drainage area; colour in mg Pt/l; CaMg (hardness) in meq/l; totP in µg/l.

4. Various lake morphometrical parameters: Dm = mean depth in m, Q = theoretical water discharge in m^3/sec, theoretical water retention time, Tw in years, percentage of the lake bed dominated by accumulation processes and fine sediments, A-areas in % of lake area, and dynamic ratio, DR (= $\sqrt{Area}/Dm$).

5. Different parameters describing the catchment area: Rock% is the percentage of bare rocks in the watershed, ADA = the area of the drainage area in m^2, RDA = the relief of the catchment area, Open land% is the percentage of open (= cultivated land).

It is evident that the target y-variable, Cs_{pi88}, may be related to the fallout (Cs_{soil}) after the Chernobyl accident, and several of the given water variables could, potentially, influence the bioavailability and biouptake of radiocesium as well as the biomasses, and hence the concentration of radiocesium in the biomasses. The morphometrical parameters could, potentially, influence the retention of radiocesium in lakes, the resuspension and the internal loading of Cs, and the catchment parameters could all, potentially, influence the runoff of cesium from land to water, i.e., the secondary load of radiocesium to the lakes. But, as stressed, all these factors cannot be of equal importance in predicting Cs_{pi88}. One simple way of quantitatively ranking such dependencies is to make a correlation analysis (Table 1.1A). It should be noted that some of the potential factors appear with high r-values vs Cs_{pi88}, like Cs_{wa87} (r = 0.88), total-P (r = 0.48), dynamic ratio (r = -0.64) and Rock% (r = 0.40), and some with low r-values. We should also remember that many, if not all, of the variables are related to one another. High correlations exist between K-concentration, conductivity, hardness and alkalinity (r > 0.9; see Håkanson, 1997). Such inter-related variables can replace one another in empirical models without causing any major loss in predictive power.

Note also that there exist great differences in the representativeness and reliability of these potential model variables. All water chemical variables vary with time and sampling location in a lake. Characteristic CV-values (coefficient of variation, CV = MV/SD, MV = mean value, SD = the standard deviation) for many important variables for this radiocesium model, like radiocesium concentrations in fish, water and sediments (from Håkanson, 1998a), and many water chemical variables (from Håkanson and Peters, 1995) are listed in Table 1.2. Many of these CV-values will be used in a following section in uncertainty tests using Monte Carlo simulations. It is evident that each different model variable (x) can be determined with different degrees of uncertainty. Morphometric parameters can often be determined very accurately (see Pilesjö et al., 1991); some model variables, like rates and distributions coefficients, on the other hand, cannot be empirically determined at all for real ecosystems, but have to be estimated from laboratory tests or theoretical derivations. This means that the values used for such model variables are often very uncertain. About 70% of the data in the frequency distribution fall within the values given by MV±SD, and 95% of the data are likely to fall within MV±2·SD. Table 1.2 shows that model variables like rates and distribution coefficients generally can be given CV-values of 0.5. The highest expected CV-values appear for certain sedimentological variables, such as concentrations in sediments.

CV-values expressed in this manner are by far the most commonly expressed statistic for this purpose in the radioecological literature (see Whicker, 1997). When more data become available, this assumption should be challenged, since it is well-known (see, e.g., Håkanson and Peters, 1995) that many water chemical variables are not normally but log-normally distributed. For non-normally distributed variables alter-

Table 1.2. Compilation of characteristic CV-values for different types of lake variables. All CV-values, except for the model variables, emanate from empirical measurements. Modified from Håkanson (1999).

	CV		CV
Catchment variables		**Lake management variables**	
Catchment area (ADA)	0.01	Secchi depth (Sec)	0.15
Percent outflow areas (OA)	0.05	Chlorophyll-a concentration (Chl)	0.25
Fallout of radiocesium (Cs_{soil})	0.10	Hg-concentration in fish muscle	0.25
Mean soil type or permeabilty factor (SPF)	0.25	Cs-concentration in fish	0.22
Lake variables		Cs-concentration in water	0.30
Lake area (Area)	0.01	Cs-concentration in sediments	0.60
Mean depth (Dm)	0.01	**Climatological variables**	
Volume (Vol)	0.01	Annual runoff rates	0.10
Maximum depth Dmax)	0.01	Annual precipitation	0.10
Theoretical water retention time (Tw)	0.10	Temperatures	0.10
Water chemical variables		**Model variables**	
pH	0.05	Fall velocities	0.50
Conductivity (cond)	0.10	Age of A-sediments	0.50
Ca-concentration (Ca)	0.12	Age of ET-sediments	0.50
Hardness (CaMg)	0.12	Diffusion rates	0.50
K-concentration (K)	0.20	Retention rates	0.50
Colour (col)	0.20	Bioconcentration factors	0.50
Fe-concentration (Fe)	0.25	Feed habit coefficients	0.50
Total-P concentration (TP)	0.35	Distribution (=partition) coefficients	0.50
Alkalinity (alk)	0.35		
Sedimentological variables			
Percent ET-areas (ET)	0.05		
Suspended particulate matter conc. (SPM)	0.20		
Mean water content for E-areas	0.30		
Mean water content for T-areas	0.20		
Mean water content for A-areas	0.05		
Mean bulk density for E-areas	0.10		
Mean bulk density for T-areas	0.10		
Mean bulk density for A-areas	0.02		
Mean organic content for E-areas	0.50		
Mean organic content for T-areas	0.50		
Mean organic content for A-areas	0.10		
Mean TP-conc. for E-areas	0.50		
Mean TP-conc. for T-areas	0.75		

native methods (see Gilbert, 1987) of expressing the relative variability could be used, such as the geometric standard deviation, the range and/or the ratio between maximum and minimum values. However, for the present purpose, this simplification seems justified. It is not likely that other statistical measures of uncertainty will substantially change the general conclusions about empirical uncertainties for radiocesium in water, sediments and biota given in this paper.

The next example in Table 1.1 illustrates the result of a stepwise multiple regression using Cs_{pi88} as target y-variable and the variables in Table 1.1A as x-variables. This is a real ranking of the different x-variables in relation to the y-variable. Note that the concentration of ^{137}Cs in water in 1987 is the most important x-variable. This is a typical collective variable which explains statistically approximately 78% (r^2 = 0.78) of the variability in the y-variable (Cs_{pi88}) among these 14 Swedish lakes. The next factor is the potassium concentration of the lake water (the mean annual K-value for 1987 is used). At the second step, the r^2-value has increased to 0.89. The third x-variable is the Open Land percentage (OL%, a measure of the cultivated land) of the catchment. Accounting for OL% increases r^2 to 0.92. The fourth and last x-variable is lake total-P concentration (mean value for 1987). It increases r^2 to 0.93.

It should be noted that this empirical model only applies for a given year after the Chernobyl accident. The most important water chemical variable in this empirical model is the lake K-concentration - the more ions similar to Cs, like K, the lower the uptake of ^{137}Cs - a case of "chemical dilution" (Black, 1957; Fleishman, 1963; Carlsson, 1978). For a "peak emission" like this, the biouptake of ^{137}Cs, and the Cs-concentration in fish, is also lower in lakes with fast water turnover than in lakes with slow water turnover - a case of normal "water dilution". The higher the lake TP-concentration, the higher the bioproduction, the greater the biomasses and the lower the Cs-concentrations in the biomasses - a case of "biological dilution".

Radiocesium seems to have a considerable particle affinity (Broberg and Andersson, 1989; Riise et al., 1990; Madruga and Cremers, 1997; Konitzer and Meili, 1997; Konoplev et al., 1997). This is very important and means that the behaviour of radiocesium in lakes is strongly influenced by the chemical and physical properties of the carrier particles (illite clays). In addition, the retention in a lake will be governed by the turnover time of the lake water; i.e., the longer the water turnover time, the larger the proportion of the initial cesium load retained within the lake (mostly in the sediments).

These are statistical results, and a central part in the derivation of the following model is to account for the mechanistic principles expressed by the empirical results given in Table 1.1. The empirical model is static, i.e., it is only valid for the Cs-concentration in pike caught in 1988 two years after the Chernobyl fallout. The following model is dynamic, i.e., it provides time-dependent predictions not just for pike, but for many species of fish with different feeding habits, for many different types of lakes and for different fallout scenarios.

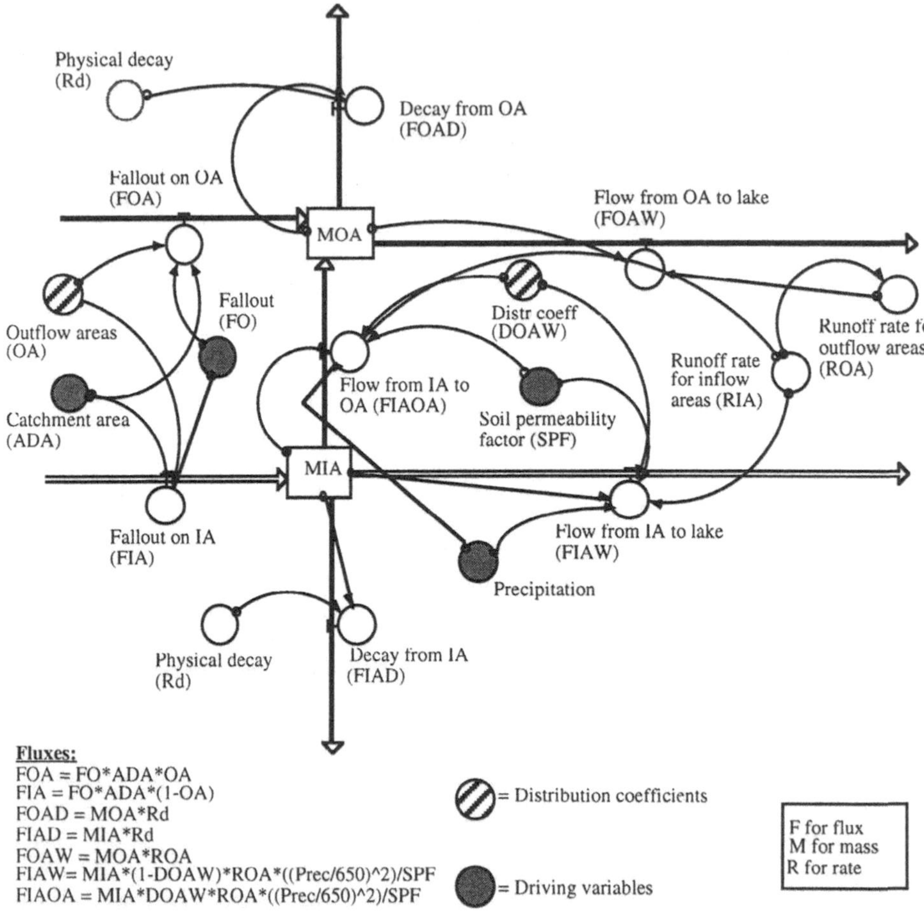

Fig. 1.4. An outline of the catchment area sub-model.

1.3. The Model - Equations

1.3.1. THE CATCHMENT SUB-MODEL - INFLOW TO THE LAKE

The catchment sub-model (see Figure 1.4) handles the inflow to the lake via the tributaries. The direct deposition (fallout multiplied by area) is the net sum of dry and wet fallout onto the catchment. The tributary load is a function of the amount of ^{137}Cs in the catchment multiplied by an outflow (transfer or retention) rate.

The catchment area is divided into two parts:

1. Outflow areas (≈ wetlands) dominated by a relatively fast turnover of substances and horizontal (land overflow) transport processes (see Eriksson, 1974; Nyström, 1985; Grip and Rodhe, 1985; Rodhe, 1987).

2. Inflow areas (≈ dry land) dominated by vertical transport processes, first through the soil horizons, then ground water transport, and, finally, tributary transport to the lake.

1.3.1.1. Outflow areas

In the following presentation, the equations and the presuppositions for the equations are given first; then there will be selected simulations to illustrate how the equations actually work.

The approach for calculating the secondary input to the lake from outflow areas (Bq per month) is given by:

$$M_{OA}(t) = M_{OA}(t - dt) + (F_{OA} + F_{IAOA} - F_{OAW} - F_{OAD})\cdot dt \quad (1.1)$$

where
$M_{OA}(t)$ = mass (= amount) of ^{137}Cs in outflow areas at time t (Bq);
F_{OA} = deposition on outflow areas (Bq per time unit); the time unit, dt, in this model for radiocesium is set to one month to get seasonal variations;
F_{IAOA} = flow from inflow areas to outflow areas (Bq per month);
F_{OAW} = flow from outflow areas to the lake (Bq per month);
F_{OAD} = physical decay of radiocesium from outflow areas (Bq per month).

The deposition on outflow areas is given by:

$$F_{OA} = FO\cdot ADA\cdot OA \quad (1.2)$$

where
FO = fallout (= deposition) in (Bq/m^2);
ADA = area of drainage area (m^2);
OA = fraction of outflow areas in the catchment (a dimensionless distribution coefficient).

It is evident that the fallout is very important for all subsequent predictions of Cs-concentrations in water and biota, and that the area of the catchment (ADA) and the fraction of outflow areas (OAs) are key factors in the secondary transport of radiocesium from land to water. Figure 1.5 illustrates how the model predicts total Cs-concentrations in water (Figure 1.5A) and in small (10 g) perch (Figure 1.5B) in lake 2217 (Hamstasjön, Sweden, one of the validation lakes in Table 1.4). The value for OA is important - the larger the outflow area, the higher the secondary transport of radiocesium from land to lake. For Nordic catchment areas, OA normally varies around 0.2; values higher than 0.3 are rare (Håkanson and Peters, 1995). In the simulations given in Figure 1.5, OA is varied from 0 to 0.3 while all else is constant. OA can be determined relatively well (from the area of bogs, mires, upstream lakes, etc.).

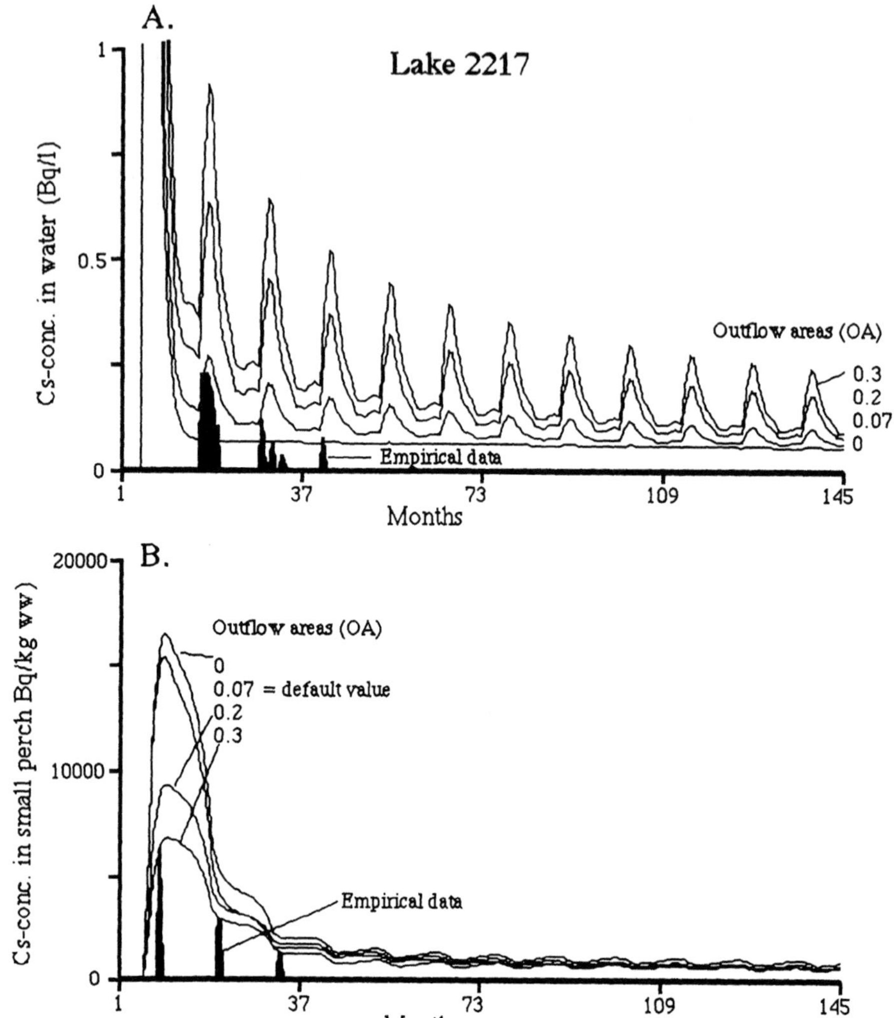

Fig. 1.5. A comparison between empirical data and modelled values for (A) total Cs-concentrations in water and (B) Cs-concentrations in 10 g perch in Lake Hamstasjön (2217) in a scenario where the value for the outflow areas of the catchment (i.e., the wet land, OA) has been changed (0, 0.07, which is the actual value for this lake, 0.2 and 0.3, which is a very high value for a catchment area).

The characteristic CV-value for OA is 0.1 (see Table 1.2). Figure 1.5 shows that there is very good correspondence between empirical data and modelled values. The value for OA for lake 2217 is 0.07. Note that the results in Figure 1.5 are from a validation, i.e., an independent test, so there has been no tuning of the model to achieve these predictions (i.e., no changes in the model constants).

The approach for calculating the retention rate regulating the flow from inflow areas to outflow areas, F_{IAOA}, is based on a default assumption that the retention rate for radiocesium in inflow areas is longer than in outflow areas. How much longer is given by the soil permeability factor (SPF). Unrealistically large fluxes will be obtained if this rate is set too high, and too low fluxes to the lake if the rate is set very low. There are also two main approaches for calculating seasonal variations in water runoff and/or tributary water discharge, which will be explained in the following text.

$$F_{IAOA} = D_{OAW} \cdot M_{IA} \cdot R_{IA} \tag{1.3}$$

where

D_{OAW} = the distribution coefficient used to partition the fluxes from inflow areas into outflow areas (OAs) and/or directly into lake water (W). The default value of D_{OAW} is 0.5, i.e., 50% of the transport goes directly to the lake water.

M_{IA} = mass (= amount) of radiocesium in inflow areas (Bq).

R_{IA} = the runoff rate for inflow areas, which is given by:

$$R_{IA} = (R_{Rd}/12) \cdot (Prec/650)^2/SPF \tag{1.4}$$

Thus, to calculate the flux from inflow areas to outflow areas, one needs (1) a default runoff rate, R_{Rd}, (2) a precipitation factor (Prec = the mean annual precipitation in mm) and (3) a soil permeability factor (SPF; see below for defnition).

The default runoff rate R_{Rd} is set to 0.04 for the first year after the Chernobyl fallout. This value emanates from studies presented by Håkanson et al. (1988). This gives the factor $R_{Rd}/12$ or 0.04/12. The runoff of radiocesium also depends on the precipitation; the more it rains, the greater the transport from land to water of radiocesium. This is handled by the dimensionless moderator $(Prec/650)^2$; 650 mm/yr is the reference value for the mean annual precipitation for European catchments/lakes and Prec is the actual precipitation in mm/yr for the given catchment. The exponent 2 means that there is a non-liner relationship between precipitation and runoff. If the actual precipitation markedly departs from the normal value, the soils will be saturated and the runoff increase much more than in unsaturated soils.

The soil permeability factor (SPF) is a dimensionless factor accounting for the fact that soils of different grain size characteristics have different retention rates for radiocesium: the more permeable, the larger the retention rate. The following default values are used for SPF:

1 for very permeable morainic and organic soils;
10 for mixed morainic soils and sandy soils;
50 for silty morainic soils;
60 for silty soils;
80 for clay soils.

If there are no data available on characteristic catchment area soil type, a SPF-value of 25 may be used as a general default value.

For simplicity, this model has no distribution coefficient for soil types (i.e., no soil Kd). The soil permeability factor (SPF) may be regarded as a collective parameter expressing many complicated relationships regulating the retention and fixation of radiocesium in soils.

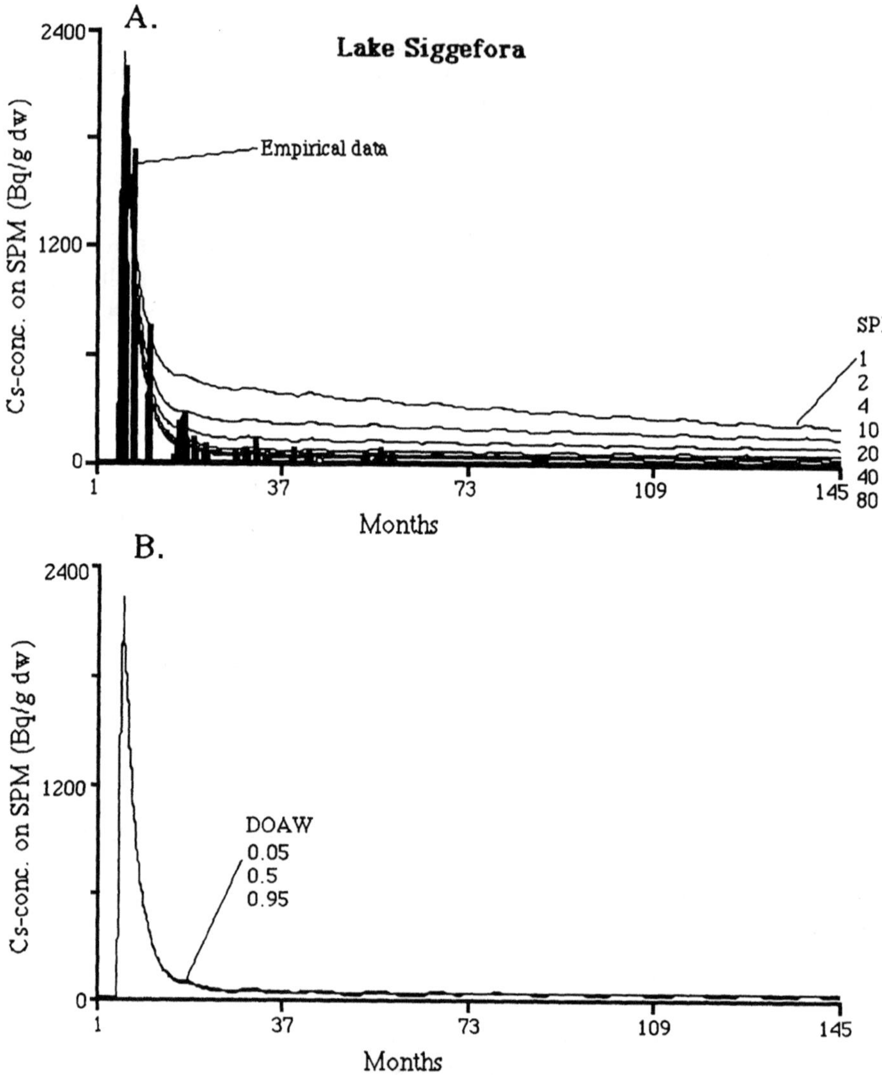

Fig. 1.6. A comparison between empirical data and modelled values for total Cs-concentrations on suspended particulate matter in Lake Siggefora in a scenario where the value for (A) the soil permeability factor (SPF) has been changed (from 1, very permeable morainic and organic soils to 80, clay soils, which is also the actual value for this catchment area), and for (B) the distribution coefficient regulating the flow from inflow areas to either outflow areas or directly to the lake (D_{OAW}; the default value for the model is 0.5).

Figure 1.6A shows that the value selected for the soil permeability factor (SPF) is important for the transport of radiocesium from land to lake, and hence also for the Cs-concentrations in water and biota and for the Cs-concentration on suspended particulate matter (SPM). The higher the soil permeability factor, the smaller the SPF-value, the faster the Cs-transport from the catchment to the lake, and the higher the Cs-concentrations on SPM in the lake. The prediction in Figure 1.6A concerns Lake

Siggefora, Sweden, one of the validation lakes (Table 1.4). Note the excellent fit between empirical data and modelled values for the actual SPF-value of 80 for this lake. This is a validation, not a calibration. Figure 1.6B gives a similar sensitivity analysis for different values of D_{OAW}, i.e., the distribution coefficient between regulating the flow from inflow areas to either outflow areas or the lake water. These predictions are not very sensitive to the choice of value for D_{OAW}, and the reason for this is that it does not matter very much if one directs one flux directly to the lake from inflow areas or redirects the flux via the outflow area to the lake. From a mechanistic point of view, however, it would be unrealistic to assume that all radiocesium would be transported only from inflow areas to one compartment, and this model structure is likely to yield the most stable results. If, however, model simplifications are requested, one could omit this distribution coefficient.

It should also be noted that there is no time-dependent function for the runoff rate from inflow areas as there is for outflow areas (see eq. 1.6). The rationale for this is that the flux of radiocesium from the inflow areas can be regarded as a slow-moving front. A certain fraction of the amount of radiocesium from the inflow areas is transported either to outflow areas or directly to the lake per time unit. The transport from outflow areas, on the other hand, is much quicker. Under default conditions, it is 25 times quicker than the transport from the inflow areas (i.e., for SPF = 25). Since the transport from outflow areas is relatively fast, the most mobile fractions of radiocesium are successively depleted from the outflow areas. This is handled by the time-dependent function $(1/(\text{time}+1)^{0.5}$. Eight months after the fallout, the time-dependent rate is 1/3 [= $1/(8+1)^{0.5}$] of the initial rate; 99 months after the fallout, the rate is 1/10 [= $1/(99+1)^{0.5}$] of the initial value.

The flux from outflow areas to lake water, F_{OAW}, (Bq/month), is given by:

$$F_{OAW} = M_{OA} \cdot R_{OA} \tag{1.5}$$

where
R_{OA} = the retention rate of ^{137}Cs in outflow areas, which in turn is calculated from (see Håkanson et al., 1996a):

$$R_{OA} = Y_Q \cdot (R_{Rd}/12) \cdot (1/\sqrt{(\text{time}+1)}) \tag{1.6}$$

time = fallout month (1 is January). The actual value of the rate is modified by the seasonal moderator for water runoff or water discharge (Y_Q, the moderator, gives an increased transport of radiocesium from land to lake during peaks in water flow). The rate is time-dependent. It decreases with time from the month of the fallout (given by $1/\sqrt{\text{time}}$). A delay function initiates the runoff to the month of the fallout plus one month (time+1). The equation will cause the initial outflow rate (0.04/12) to decrease with time, and the seasonal moderator for Q will create a seasonal variability in the outflow rate.

Figure 1.7A shows that the value selected for the runoff rate (R_{Rd}) is rather important for the transport of radiocesium from land to lake, and hence also for the Cs-concentrations in water and biota and for the Cs-concentration on suspended particulate matter (SPM). The higher the rate, the greater the Cs-transport from the catchment, and

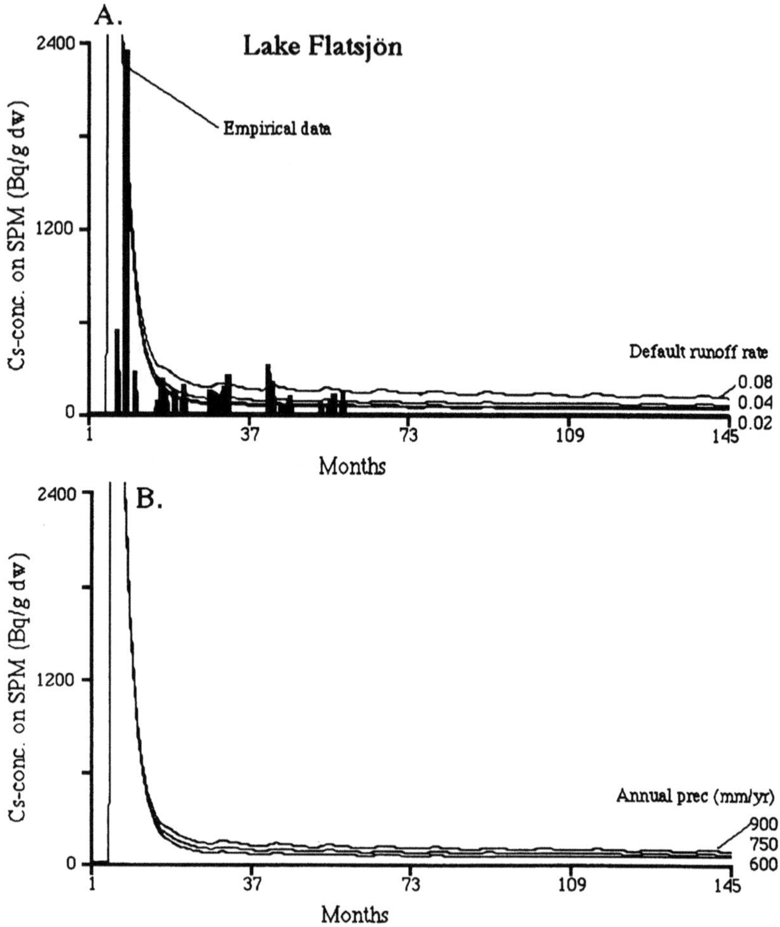

Fig. 1.7. A comparison between empirical data and modelled values for total Cs-concentrations on suspended particulate matter in Lake Flatsjön in a scenario where the value for (A) the default runoff rate (R_{Rd}) has been changed (0.02, 0.04 (default value) and 0.08), and for (B) the mean annual precipitation (600, 750, which is the actual value for this lake, to 900 mm/yr).

the higher the Cs-concentrations on SPM in the lake. The prediction in Figure 1.7A concerns Lake Flatsjön, one of the calibration lakes (Table 1.3). Also note the fine fit between empirical data and modelled values for the default R_{Rd}-value of 0.04 for this lake. This is a calibration. However, there has been no tuning of the model. All model variables have been kept at the generic default values, and only the lake-specific variables have been changed. Figure 1.7B gives a similar sensitivity analysis for values of the mean annual precipitation (= prec). The predictions are not sensitive to the choice of the prec-value.

Table 1.3. Data on the lakes used for calibration the radiocesium model. 1) Marcus Sundbom, pers. comm., 2) IAEA (1999), 3) Jim Smith, pers. comm., 4) Håkanson (1991).

A.

Lake		**Fallout** (kBq/m^2)	**Area** (km^2)	**Dm** (m)	**Dmax** (m)	**pH**
Flatsjön	Sweden	70	0.61	2.5	4	8
Hillesjön	Sweden	100	1.63	1.7	3.1	7.3
Devoke Water	England	17	0.34	4	15	6.5
Constance	Switzerl.	6	537.4	90	252	7
Fäbodsjön (2102)	Sweden	4	0.35	1.1	2.3	6.3
Ecklingen (2110)	Sweden	3	0.44	4.7	19.2	6.6
Långsjön (2120)	Sweden	4	0.16	2.1	6.2	6.2
Selasjön (2201)	Sweden	70	0.25	8.3	16.3	5.2
Huljesjön (2204)	Sweden	35	1.26	8.5	26.2	6.2

B.

Lake	**TP** (µg/l)	**Tw** (yr)	**Data on**	**Reference**
Flatsjön	20	0.94	Susp., perch, roach, pike	1
Hillesjön	20	0.36	Water, perch, roach, pike	2
Devoke Water	10	0.24	Water, perch, trout	2
Constance	10	4.2	Water	3
Fäbodsjön (2102)	26.4	0.02	Water, perch, pike	4
Ecklingen (2110)	11.8	0.48	Water, perch, pike	4
Långsjön (2120)	8.5	0.3	Water, perch, pike	4
Selasjön (2201)	9.5	0.33	Water, perch, pike	4
Huljesjön (2204)	8.3	0.65	Water, perch, pike	4

The month of the fallout is important, not so much for the total Cs-concentrations in water, but certainly for the Cs-concentrations in biota. One example of this is given in Figure 1.8, which illustrates how the model predicts the total Cs-concentrations in water (Figure 1.8A) and in 12.5 g perch in lake 2110 (Lake Ecklingen, Sweden, one of the calibration lakes; see Table 1.3). The critical point in this context is when fallout happens relative to (1) the ice-conditions, (2) the seasonal characteristics of the tributary water discharge and (3) the seasonal variation in the bioproduction. These conditions vary among and within lakes, e.g., with latitude, altitude and longitude. It should be noted that there is a very good correspondence between empirical data and modelled values in Figure 1.8. It should also be noted that there has been no tuning of the model to achieve the predictions given in Figure 1.8. All model variables have been kept at the generic default values, and only the lake-specific variables have been changed.

Y_Q = seasonal moderator for the water discharge (= Q, dim. less); Y_Q is defined either as the ratio between the mean monthly surface runoff rate (MMSR) and the mean annual surface runoff rate (MASR), i.e., Y_Q = MMSR/MASR, or from the ratio

Table 1.4. Data on the lakes used for validating the radiocesium model. 1) Marcus Sundbom, pers. comm., 2) IAEA (1999), 3) Jim Smith, pers. comm., 4) Håkanson (1991).

A.

Lake	**County**	**Fallout** (kBq/m^2)	**Area** (km^2)	**Dm** (m)	**Dmax** (m)	**pH**	**TP** (µg/l)
IJsselmeer	Netherl.	2.2	1147	4.3	10	8.5	60
Iso Valkjärvi	Finland	70	0.042	3.1	8	5.1	11
Ø. Heimdalsv.	Norway	130	0.78	4.7	20	6.8	10
Esthwaite	England	2	1	6.4	15	8	25
Siggefora	Sweden	30	0.73	4.2	11	7.2	10
Hamstasj. (2217)	Sweden	50	0.18	3.8	7.8	6.6	22.9
Lill-Selasj. (2214)	Sweden	70	0.07	2.9	9.2	5.1	10.9
Zürich	Switzerl.	5.9	537.4	90	252	7	10

B.

Lake	**Tw** (yr)	**Trophic level**	**Fish species**	**Abiotic var.**	**Reference**
IJsselmeer	0.41	Hypertr.	Roach, smelt, perch	Water	2
Iso Valkjärvi	3.0	Oligotr.	Pike, whitef., perch	Water	2
Ø. Heimdalsv.	0.17	Oligotr.	Minnow, trout	-	2
Esthwaite	0.19	Eutr.	-	Water	3
Siggefora	0.78	Oligotr.	Pike, roach, perch	SPM	1
Hamstasj. (2217)	0.21	Mesotr.	Perch	Water	4
Lill-Selasj. (2214)	0.02	Oligotr.	-	Water	4
Zürich	4.2	Oligotr.	-	Water	3

between the mean monthly precipitation and the mean annual precipitation, or from the mean monthly water discharge and the mean annual water discharge. This is illustrated in Figure 1.9 (from Abrahamsson and Håkanson, 1998).

Climatic variables, like river discharge and water temperature, influence many important processes in a lake ecosystem. For example, the retention time of the water and, hence, the retention of substances in lake water, are regulated to a large extent by water discharge in tributary rivers. Differences in hydrological regime also, evidently, affect the transport of nutrients and toxic substances from catchments (Brittain et al., 1994).

Since river discharge depends on many more or less stochastic processes, and has a high degree of variability between years for a given river, it is very difficult to give a reliable prediction of the discharge (Q) for a specific river at a given time. A

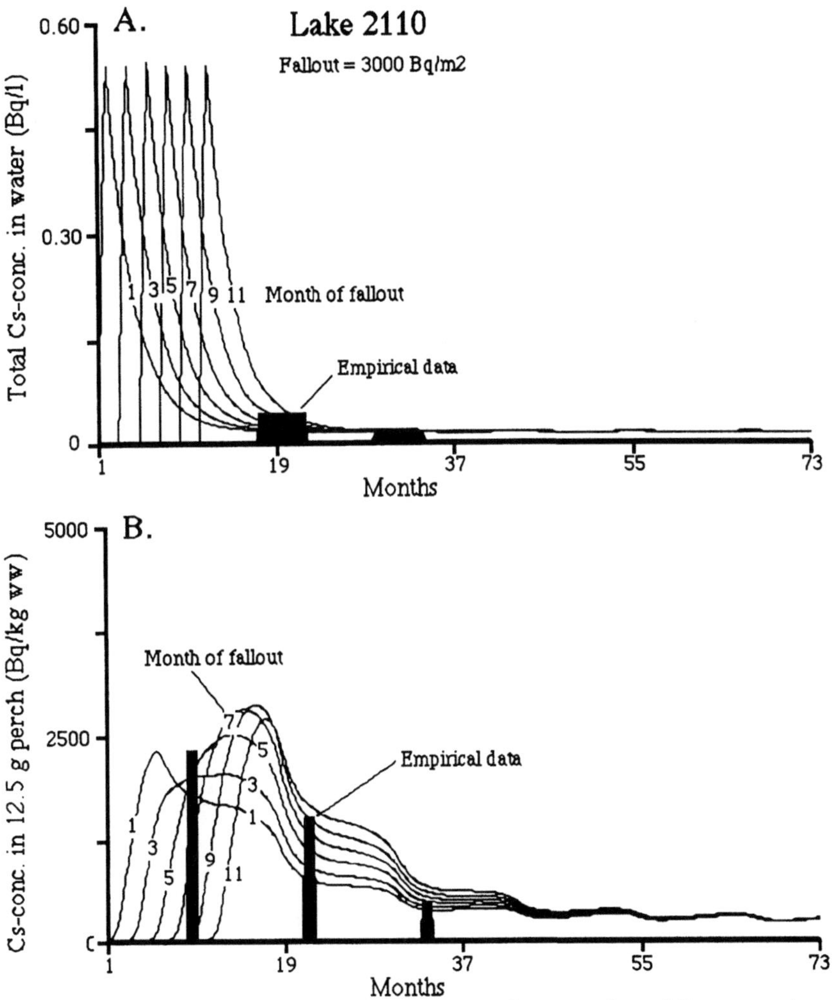

Fig. 1.8. A comparison between (A) empirical data and modelled values for total Cs-concentrations in water and (B) Cs-concentrations in 12.5 g perch in Lake Ecklingen (2110) in a scenario when the month of the fallout has been changed (fallout, in January, March, May, July, September and November).

standard procedure is to measure the river discharge for a long period of time (decades) and then give a statistical estimate of a probability that Q is going to be within a certain range at a certain time. That method is appropriate for many purposes, providing that a sufficiently long and reliable set of empirical data is available (Chow, 1988). However, if empirical data on Q are not available, which is certainly the case for a very large number of rivers, other methods are necessary, e.g., statistical/empirical methods to predict Q from, e.g., soil type distributions, vegetation types, etc. Such models (e.g., HFAM by Hydrocomp Inc.) can be very precise and valuable, but they require a lot of

proper value for mean monthly Q, which is the target in this sub-model. This sub-model is meant to yield predictions for Q, which can be accepted in ecosystem models for toxic substances where the focus is on the predictive power for the Cs-concentrations in water, sediments and biota. This is the main reason why we have tried to develop a model that calculates monthly values for river discharge based on readily available map parameters, such as latitude, altitude and precipitation. But we also think that this model could be useful in a wider context of ecological modelling when river discharge and/or lake water retention time are important modelling variables, i.e., when the target variables being predicted are biological (e.g., fish biomasses) and/or chemical (e.g., lake pH and lake phosphorus concentration) and where the inevitable uncertainties in the predicted values of water discharge associated with this simple model can be accepted. Examples of such models are lake liming models (Ottosson and Håkanson, 1997), and models for evaluating effects of fish-farm emissions on lakes (Håkanson et al., 1998).

To calibrate and validate the new Q-model, an extensive data set from more than 200 European rivers was used. The discharge of the chosen rivers is not affected by regulation for hydropower or irrigation purposes since that produces unnatural seasonal flow patterns. The river discharge data for the Swedish rivers were recorded by the Swedish Meteorological and Hydrological Institute (SMHI, 1995a-d). The data set for the other rivers comes from UNESCO (1993). The time series for the monthly data are at least six years long, and some of them are as long as 80 years. The data sets have been divided into two parts of equal size - one for the calibration and one for the following validation. The rivers used in the validation have characteristics in approximately the same range as the rivers used in the calibration concerning the relevant parameters, such as longitude, latitude and altitude. Figure 1.9 shows that the only obligatory driving variables for this Q-model are altitude, latitude, mean annual precipitation and catchment area.

To simulate the monthly variations in river discharge, six seasonal variability norms are utilised for Europe (see Figure 1.9). A seasonal variability norm is used to add a seasonal pattern to an annual value (Håkanson, 1996). Two of these norms should represent the typical seasonal flow pattern in the most southern and northern parts of Europe, respectively. Two other norms should describe the effect of altitude on monthly variability in river discharge and two should represent the typical flow pattern of rivers with very small and very large mean annual discharges (Q_A). Depending on the location and the mean annual discharge of the specific river, the six seasonal variability norms are weighted together and a site-specific seasonal variability norm is calculated. In the calibration, it became obvious that the addition of longitude did not significantly increase the degree of explanation of the model and longitude was, therefore, excluded.

To quantitatively account for how latitude (Lat), altitude (Alt), and mean annual discharge (Q_A) for a specific river influence the seasonal (monthly) variability, different weighting factors ranging from zero to one were developed. For example, the weight factor for latitude should be zero for a river in the southernmost part of Europe (35°N) and one in the northernmost part (70°N). To account for the fact that the relationships do not have to be linear, each weighting factor was given an exponent. The exponents are used to control in which range the changes in the parameters are most critical. This is, of course, still a very simple approach for simulating the influence of different parameters. This approach gave the equation for the seasonal moderator for Q (Y_Q) given in Figure 1.9. The values for the six norms are also given in Figure 1.9.

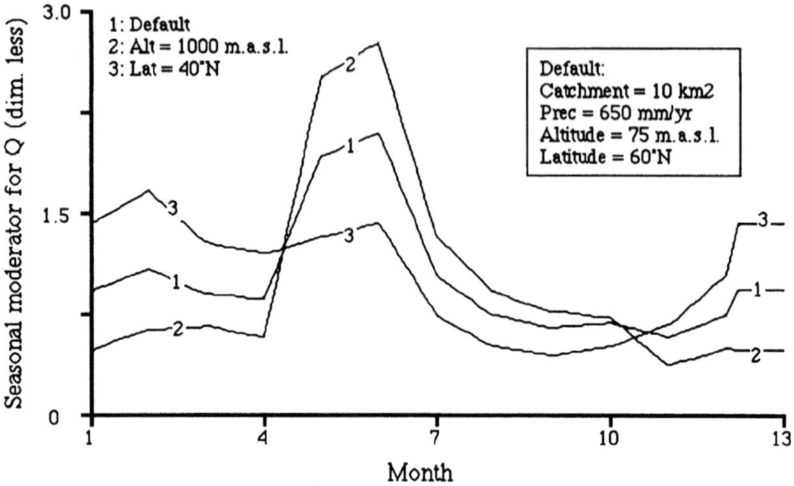

Fig. 1.10. Predicted seasonal patterns in tributary water discharge (Q), as expressed by the dimensionless moderator, Y_Q, under defined default conditions (curve 1), for a catchment at altitude 1000 m.a.s.l. (curve 2) and for a catchment at latitude 40 °N (curve 3).

In spite of the fact that river discharge is a variable with great temporal and spatial variability, this sub-model has yielded good predictions of monthly average Q based on only very few and readily accessible map parameters. The best results of the validation were achieved for the rivers with a mean annual discharge in the range 1-500 m^3/s. More uncertain predictions are obtained for the smallest and largest rivers. The range is, however, large enough to include most European rivers (88 of 114 in the calibration and 90 of 119 in the validation). So, given that limitation, the model certainly has a wide range of applicability.

Figure 1.10 gives a test to illustrate how differences in altitude and latitude will influence the seasonal moderator for Q. The default conditions are given by a lake with a catchment area of 10 km^2, a mean annual precipitation of 650 mm/yr, an altitude of 75 m.a.s.l. and a latitude of 60°N. Curve 1 in Figure 1.10 gives the characteristic seasonal variations in Y_Q. If a similar lake is situated at an altitude of 1000 m.a.s.l., it is likely that the precipitation is more unevenly distributed over the year, and the seasonal variability in Q larger. If the lake is placed at latitude 40°N, there is a less pronounced seasonal variability pattern in Y_Q. Note that this is based on extensive calibrations and validations based on empirical Q-series from many European catchment areas.

The physical decay from outflow areas (F_{OAD}) is defined as:

$$F_{OAD} = M_{OA} \cdot R_d \qquad (1.7)$$

In radioactive decay, a constant fraction of the remaining activity is lost per unit time and so the activity at any time t, A(t), can be calculated from the initial activity (A_0), the rate constant of decay (k) and time (t) as:

$$A(t) = A_0 \cdot e^{-k \cdot t} \qquad (1.8)$$

The physical halflife of the radionuclide is the time required for the activity to decrease by 50%. When $A(t)/A_0 = 0.5$, t is the halflife and eq. 1.8 can be rearranged as:

$$-k = \ln(0.5)/t_{0.5} \quad (1.9)$$

For example, since the physical halflife ($t_{0.5}$) of radioactive cesium is 30.2 years, eq. 1.9 may be solved for k:

$$-k = 0.693/30.2 \quad (1.10)$$

In this model, we write $-k = R_d = 0.693/(30.2 \cdot 12)$, since all calculations are done on a monthly basis.

Thus, when one calculates a halflife or its corresponding retention rate, it is often assumed that the given department (= box) can be treated as a single compartment that loses a certain fraction of itself per unit time or (since dt is set to 1 month in this model for ^{137}Cs) every month. If T is the "full" time, then 1.386 (= 0.693/0.5) from $0.693/(0.5 \cdot T)$ is often referred to as the halflife constant.

1.3.1.2. Inflow areas

The following differential equation gives all the fluxes to and from the compartment "inflow areas".

$$M_{IA}(t) = M_{IA}(t - dt) + (F_{IA} - F_{IAW} - F_{IAD} - F_{IAOA}) \cdot dt \quad (1.11)$$

where
$M_{IA}(t)$ = mass (= amount) of ^{137}Cs in inflow areas at time t (Bq);
F_{IA} = deposition on inflow areas (Bq per month);
F_{IAW} = flow from inflow areas to lake water (Bq per month);
F_{IAOA} = flow from inflow areas to outflow areas (Bq per month), as given by eq. 1.3.

The deposition (F_{IA}) on inflow areas is given by:

$$F_{IA} = FO \cdot ADA \cdot (1-OA) \quad (1.12)$$

The flow from inflow areas to lake water, F_{IAW}, is:

$$F_{IAW} = M_{IA} \cdot (1-D_{OAW}) \cdot R_{IA} \quad (1.13)$$

The physical decay from outflow areas is given by:

$$F_{OAD} = M_{OA} \cdot R_d \quad (1.14)$$

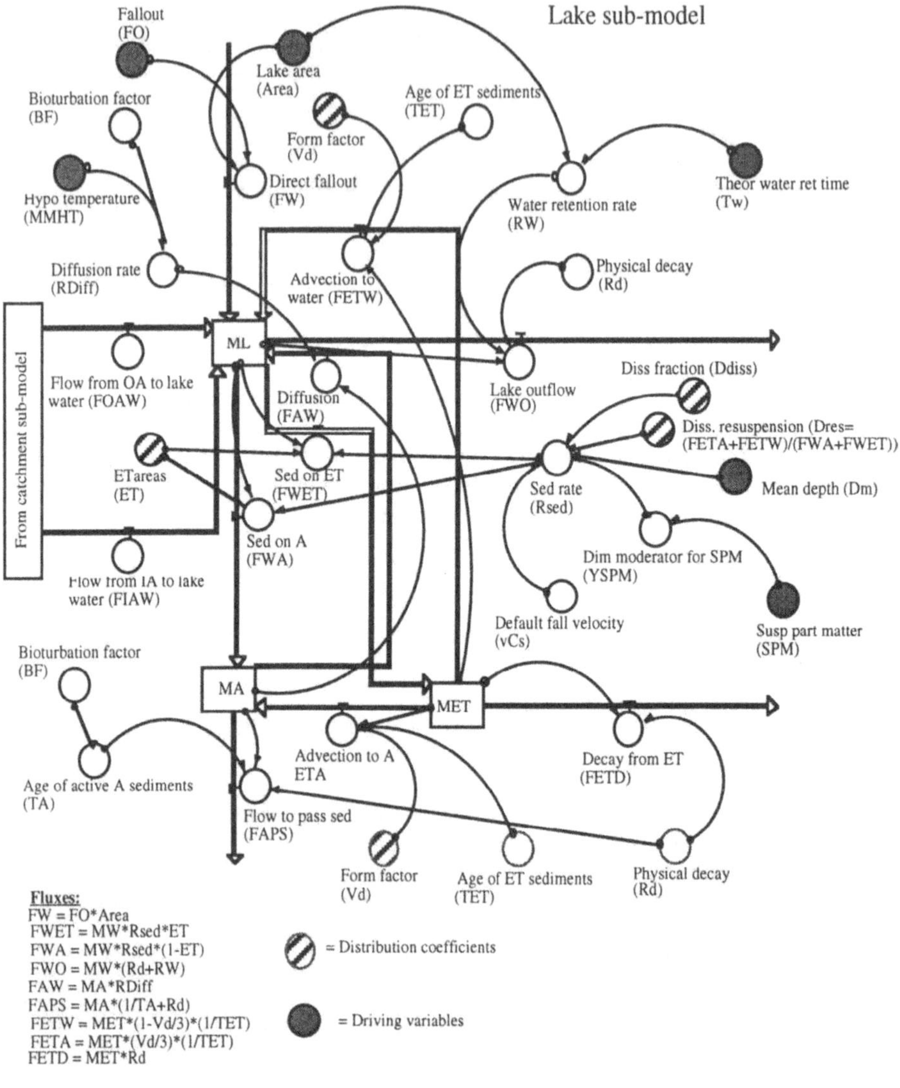

Fig. 1.11. An outline of the abiotic part of the lake model.

1.3.1.3. Direct fallout

The direct fallout to the lake (F_W; see Figure 1.11) is simply fallout, FO (e.g., Bq/m^2) multiplied by lake area (Area in m^2), i.e.:

$$F_W = FO \cdot Area \quad (1.15)$$

These equations handle all primary load functions. Next, we will focus on internal lake processes.

1.3.2. INTERNAL PROCESSES

1.3.2.1. Lake water

In this model, the entire lake volume is treated as one compartment, so there is no distinction between surface water and deep water and no stratification and/or mixing. The fluxes of radiocesium to and from the compartment "lake water" are given by the following equation:

$$M_W(t) = M_W(t - dt) + (F_W + F_{AW} + F_{OAW} + F_{IAW} + F_{ETW} - F_{WO} - F_{WA} - F_{WET}) \cdot dt \quad (1.16)$$

where
M_W = amount of ^{137}Cs in lake water (Bq);
F_{AW} = diffusion from A-areas (Bq/month);
F_{ETW} = advective transport (= resuspension) from ET-areas to lake water (Bq/month);
F_{WO} = outflow of radiocesium from lake water (Bq/month);
F_{WA} = sedimentation from lake water to A-areas (Bq/month);
F_{WET} = sedimentation from lake water to ET-areas (Bq/month).

The diffusive transport of radiocesium from A-areas to lake water, F_{AW}, is given by:

$$F_{AW} = M_A \cdot R_{Diff} \quad (1.17)$$

where R_{Diff} = the diffusion rate (1/month).

From Comans and Hockley (1992; see also Hilton, 1997; Konoplev et al., 1997), the full equation for the partitioning of ^{137}Cs in sediments is:

$$Kd = K_{c1} \cdot C_{FES}/(C_{K'} + K_{c2} \cdot C_{NH4}) \quad (1.18)$$

where
C_{FES} = the "frayed edge site" concentration;
$C_{K'}$ = the K-concentration in the interstitial sediment water;
K_{c1} = the selectivity coefficient between radiocesium and potassium;
K_{c2} = the selectivity coefficient between potassium and ammonium;
C_{NH4} = the concentration of ammonium in the sediments.

Since it is generally very difficult to access reliable empirical data for the model variables in eq. 1.18 for lake sediments, one can simplify the approach by assuming that one can use the K-concentration for the lake water also for the sediments and that the ammonium concentration can be estimated linearly from the rate of sedimentation and the hypolimnetic temperature. In that case, this sub-model does not require extra driving variables, just the driving variables already available in the model for other purposes. It is assumed (see Comans et al., 1998) that a default value for the diffusion rate for radiocesium in sediments is about 0.002 per year. The normal default value for the hypolimnetic temperature is set to 4 (°C). The diffusion rate is also set as a function of bioturbation, which influences the age of the A-sediments. If the sediments go anoxic, bioturbation is halted but the diffusion of radiocesium from sediments increases (Alberts et al., 1979). From these presuppositions, R_{Diff}, may be estimated accordingly.

If the sediments are oxic (i.e., when the Bioturbation Factor, BF, is < 1), the diffusion rate is given by:

$$R_{Diff} = (0.002/12)\cdot(GS/100)\cdot(MMHT/4) \quad (1.19)$$

where
GS = gross sedimentation on A-areas (in µg dw/cm^2·day). The normal value for GS is set to 100 µg dw/cm^2·day. The critical GS-value is set to 2000 µg dw/cm^2·day. If GS is higher than that, the sediments are likely to go anoxic and bioturbation is halted.

If the sediments are anoxic, the R_{Diff}-value is, for simplicity, set 20 times larger than under oxic conditions, as given by eq. 1.19.

The more stable the stratification, the higher the mean monthly hypolimnetic temperature (MMHT), the greater the microbiological activity, the greater the mineralization of organic matter, the greater the oxygen consumption, and the higher the potential diffusion of radiocesium from active sediments (Alberts et al., 1979). Accounting for hypolimnetic temperature in this manner means that many complex processes are lumped into one collective variable, MMHT.

There are, basically, three ways to access temperature data:

(1) From empirical measurements in the given lake. This is generally the most preferable approach. It is, however, evident that adequate sets of reliable empirical data are not available for many lakes.

(2) From standard meteorological tables. Default values used for MMHT and mean monthly epilimnetic temperature (MMET) for Nordic lakes in this approach are:

Month:	1	2	3	4	5	6	7	8	9	10	11	12
MMET:	0	0	3.0	4.0	7.0	11.0	16.0	20.0	15.0	7.0	4.0	0
MMHT:	4.0	4.0	4.0	4.0	4.5	6.0	8.0	10.0	8.0	4.0	4.0	4.0

These are meant to be typical mean monthly water temperatures in °C. With these values, it is clear that homothermal conditions prevail during April and

November. During the rest of the year, there is stratification. It is evident that this temperature regime is only valid for dimictic lakes.

(3) From predictive sub-models. The temperature sub-model illustrated in Figure 1.12 is incorporated in this model. This sub-model has been presented by Ottosson and Abrahamsson (1998). It is well known that water (and air) temperature is governed by many complicated climatological relationships. This approach accounts for:

- Altitude (Alt, in m above sea level). The higher the altitude, the lower the lake temperature and the greater the seasonal variability in temperature, if everything else is constant.
- Latitude (Lat). The higher the latitude, the lower the lake temperature and the greater the seasonal variability in temperature, if all else is constant. Latitude is given in °N.
- Continentality, or distance from ocean (in km). The idea is to have a relevant, simple measure describing the degree of continental influence on lake temperature: the farther away from the ocean, the more continental, the colder and the greater the seasonal variability in temperature, if all else is constant. The distance is given in km.
- Lake volume (V in m^3). The larger the volume, the smaller the seasonal variability in temperature.

In this approach, seasonal variability of both surface (or epilimnetic) and bottom (or hypolimnetic) water temperature will be predicted from these four map parameters. This approach uses:

1. A seasonal variability norm for surface water temperature. The monthly data constituting this norm are given in Figure 1.12. The norm is constructed to illustrate a standardised case of extreme seasonal variability in surface water temperature. Note that this norm is deliberately constructed to be unrealistically symmetric; in spite of this, one obtains realistic values of seasonal variations when the norm is being smoothed. This norm could, naturally, also be based on either empirical data from lakes with extreme seasonal variability in temperature, or on other types of theoretical curves defined for given purposes. This norm has several features. The monthly values are not dimensionless but given in °C. The dimensionless moderator will be defined by the ratio between predicted monthly mean temperatures and the predicted mean annual temperature. The range between the lowest value (-20 °C for December) and the highest value (+20 °C for June) should be high. The main point again is evidently not that this particular norm should give the most realistic description of mean monthly temperature in an extreme lake, but rather that this seasonal variability norm for temperature and an appropriate smoothing function should give realistic predictions of monthly temperatures for all lakes to which the norm is meant to apply.

2. A smoothing function based on the four map parameters is used to level out the norm. A first-order exponential smooth (SMTH) of the input (the norm for temperature) is used (see Figure 1.13).

Temperature sub-model

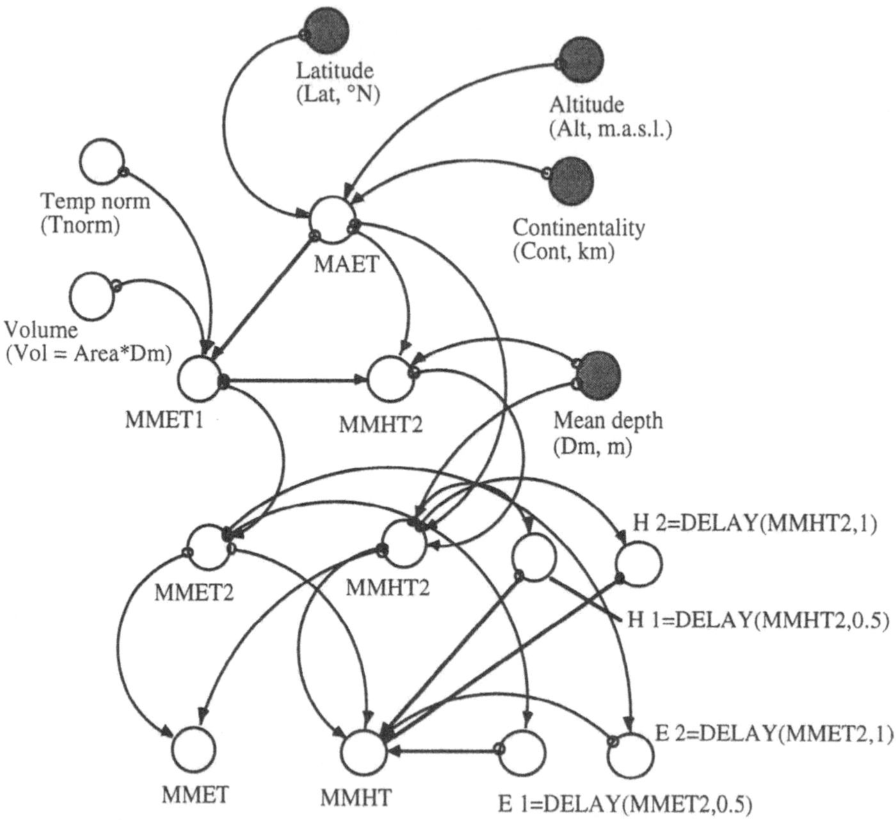

Equations:
MAET = Mean annual epilimnetic temperature, °C
MAET = 44-(750/(90-Lat^0.85))^1.29-0.1*Alt^0.5-0.25*(Cont^0.9+500)^0.52;
If MAET < 4 then 4
MMET = Mean monthly epilimnetic temperature, °C
MMET = MAET+SMTH(Tnorm,MAET/6*(Vol*10^(-6))^0.1,1)
MMHT = Mean monthly hypolimnetic temperature, °C
If MMHT1 < 4 then MMHT = 4 else if MAET < 4 or MAET > 17 or Dm < 2
then MMHT = MAET else
MMHT = SMTH(MMET1,MAET^0.5,0.51*MAET)/(0.5/(1.1/(Dm+0.1)+0.2))

Month:	1	2	3	4	5	6	7	8	9	10	11	12
Temp norm:	-8.0	-2.0	0	2.0	8.0	20.0	8.0	2.0	0	-2.0	-8.0	-20.0

Fig. 1.12. The sub-model for lake temperatures (compiled from Ottosson and Abrahamsson, 1998).

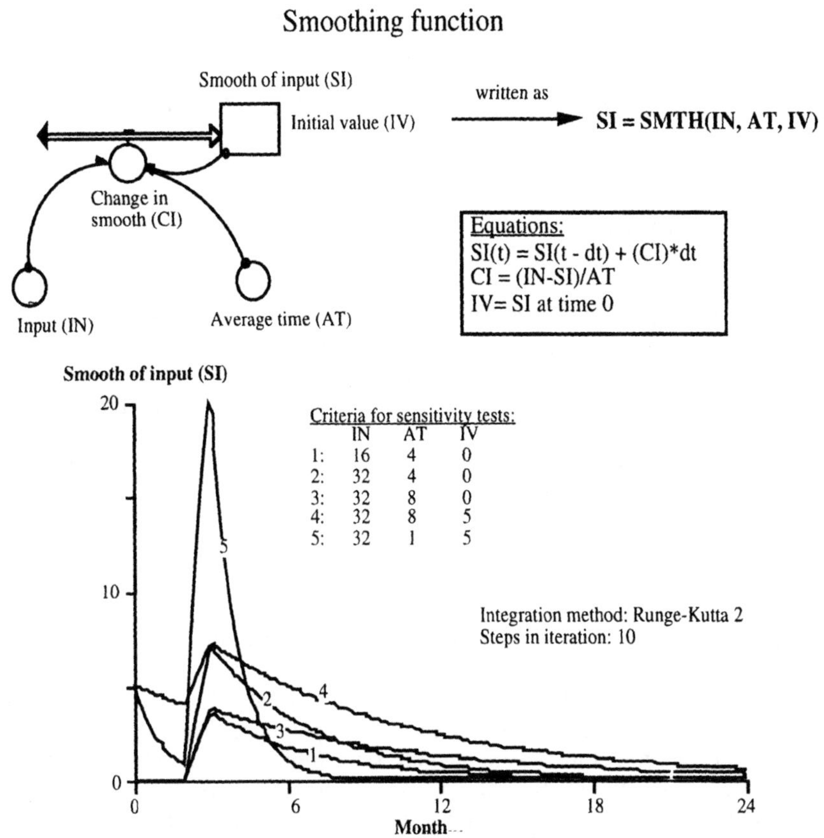

AT = 8	IV	1	2	3	4	5	6	7	8	9	10	11	12
SI	5	4,41	4,09	7,17	6,33	5,59	4,93	4,35	3,84	3,39	2,99	2,64	2,33
IN	0	0	32	0	0	0	0	0	0	0	0	0	0

Data for curve 4 for 12 time steps

Fig. 1.13. Illustration of the approach to determine the smoothing function. Modified from Håkanson (1999).

In this model, this exponential smoothing function is used in several contexts. The smoothing function is written as: SMTH(input, average time, initial value), or, using a differential equation:

$$SI(t) = SI(t-dt)+CI \cdot dt \tag{1.20}$$

where
SI = smooth of input (a time series of data);
CI = change in smooth; CI = (IN-SI)/AT, where IN = input, AT = average time.

The mean annual epilimnetic temperature, MAET (°C), is utilized as an averaging function. It is calculated from the following expression:

$$MAET = 44 - ((750/(90-Lat))^{0.85})^{1.29} - 0.1 \cdot Alt^{0.5} - 0.25 \cdot (Cont^{0.9}+500)^{0.52} \qquad (1.21)$$

where
Lat = latitude (°N);
Alt = altitude (m.a.s.l.); and
Cont = approximate distance (in km) from the lake to the ocean.

These empirical constants have been obtained after calibrations using data for a wide range of lakes (see Ottosson and Abrahamsson, 1998).

MAET should be high close to the equator (Lat = 0); the value decreases with the square root of altitude (all else being constant). MAET is rather independent of continental cooling near the oceans [given by $(Cont^{0.9}+500)^{0.52}$].

The mean monthly epilimnetic temperature (MMET) of a given lake may then be estimated from the following expression:

$$MMET = MAET + SMTH(T_{norm}, (MAET/6) \cdot ((10^{-6} \cdot Vol)^{0.1}), 1) \qquad (1.22)$$

where
T_{norm} = the norm for surface temperature (see Figure 1.12); Vol = volume (km^3).

The input is the T_{norm}. The averaging function is MAET/6. The higher the MAET, the smoother the curve for monthly surface temperatures. The factor of 6 is an empirical constant used to get a relevant smooth, since high values of MAET, e.g., 25 (°C), would produce an almost straight line, and 25/6 = 4.2 gives a more realistic smooth. The initial value is 1. This means that one first calculates MAET, then, for the summer period, positive values are added to MAET and during the winter negative values. When the lake is perennially frozen (if MAET < -10), MMET is set to 0. The size of the lake (Vol) is also used in the averaging function to smooth the norm (but Vol will not influence MAET) in such a way that lakes with a large volume will have a smaller monthly variability in surface temperature than small lakes. This is given by $Vol^{0.1}$. The exponent 0.1 is an empirical constant applied to obtain a realistic smooth.

The values of hypolimnetic temperatures (°C), naturally, can only appear in dimictic lakes, which range from latitudes of around 30 to 60 and altitudes 0 to 4000 (see Wetzel, 1983). A limit should also be set for shallow lakes with a dynamic ratio (DR = √Area/Dm; Dm = mean depth) higher than 3.8. Such lakes are not generally dimictic but polymictic (see Håkanson and Jansson, 1983). For dimictic lakes, this approach assumes that the relationship between surface and bottom temperatures depends on the mean depth: the smaller the mean depth, the smaller the difference between surface and bottom temperatures. The curve for the mean monthly hypolimnetic temperature (MMHT) should be markedly smoother than the curve for the mean monthly surface temperature (MMET). If the predicted mean annual water temperature (MAET) is higher than 17 (°C), it would probably be warm monomictic; if MAET is below about 4 (°C), the lake would be cold monomictic. Between these limits, the hypolimnetic temperatures are predicted from:

$$MMHT = SMTH(MMET, MAET^{0.5}, 0.51 \cdot MMET)/(0.5/(1.1/(Dm+0.1)+0.2))) \tag{1.23}$$

• The function to be smoothed (the input) is the predicted mean monthly surface temperature (MMET).
• The seasonal variability for the mean monthly hypolimentic temperature (MMHT) should be smaller than for MMET. This is given by the averaging value of $MAET^{0.5}$.

Figure 1.14 shows predictions for MMET and MMHT with this model, which is meant to be generally applicable to European lakes.

A. This gives the surface and bottom temperatures (MMET and MMHT) for a lake at altitude 75 m.a.s.l., latitude 60°N, continentality 100 km, area 1 km^2 and mean depth 4.3 m. Note the difference between summer and winter values. The lake circulates during spring and autumn when these two curves cross, otherwise the lake is stratified

B. Simulations for a lake at altitude 1000 m (otherwise the same conditions as in Figure 1.14A). Note that the winter stratification is stronger in this case.

C. This gives simulations for a lake at latitude 40°N. Note that the lake is not dimictic (it is warm monomictic, as it should be). It only stratifies during summer, and the temperatures are generally significantly higher, as compared to Figure 1.14A.

D. Simulations for a lake at 1000 km from the Sea. This lake is markedly dimictic.

Figure 1.15 illustrates how surface water temperatures influence the radiocesium concentrations in fish. The mechanisms for this will be discussed later, but temperature influences the biological halflife (= BHL) of ^{137}Cs in fish. The colder the water, the longer the BHL-value. The calculation in Figure 1.15 concerns lake 2201 (Selasjön, Sweden, one of the calibration lakes; see Table 1.3) and Cs-concentrations in 10 g perch and 1000 g pike. The epilimnetic temperatures are important for the Cs-concentrations in fish. In this test, all predicted default values of MMET for this lake have been changed by a factor of 1.2 and 0.8, respectively. It is worth noting the good correspondence between empirical data and modelled values. There has been no tuning of the model to achieve these predictions. All model variables have been kept at the generic default values, and only the lake-specific variables have been changed.

The following two simulations are meant to illustrate the role of the diffusion sub-model (eq. 1.19). Figure 1.16 gives a compilation and a ranking of all abiotic monthly fluxes (Bq/month) in this model for Lake Esthwaite Water, U.K. (one of the validation lakes in Table 1.4). It should be noted that there are MAJOR differences in the roles that the given processes play in this lake. Outflow from the lake, secondary inflow to the lake from outflow areas of the catchment (OA to lake), secondary inflow to the lake from inflow areas of the catchment (IA to lake) and sedimentation on A-areas (Sed A) totally dominate the fluxes, and diffusion from A-areas is the smallest flux of all.

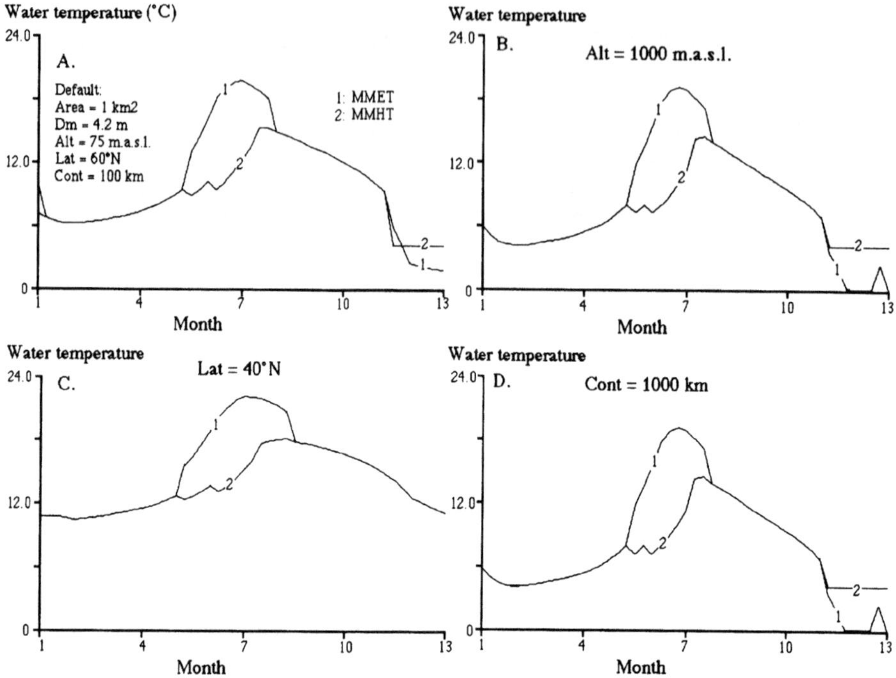

Fig. 1.14. Illustrations of the temperature sub-model.
A. Predictions in the dimictic default lake.
B. Predictions if this lake is at altitude 1000 m.a.s.l.
C. Predictions if this lake is at latitude 40 °N.
D. Predictions if this lake is 1000 km from the sea.

This means that in this lake one could omit diffusion and the predicted Cs-concentrations in water and fish, i.e., the two target variables in this model would be almost the same. However, the ranking of the fluxes is different in different lakes, so diffusion can be very important in other lakes, and even in this lake under different presuppositions. This is shown in Figure 1.16. This scenario is meant to illustrate a condition typical in many lakes and reservoirs with direct point-source emissions of radionuclides, e.g., cooling ponds. In such cases the secondary inflow from land to lake would be small or negligible, and most of the radionuclides would be found in the sediments. In the case given in Figure 1.17, all inflows to Lake Esthwaite Water from the catchment were stopped in month 13 (i.e., January of 1987). The radiocesium transported to the lake during 1986 is stored in the sediments, which for the following years would be the only source of radiocesium. Under these conditions, diffusion is one of the MAIN fluxes in this lake.

The advective transport from ET-areas to lake water, F_{ETW}, is given by:

$$F_{ETW} = (M_{ET} \cdot (1 - Vd/3))/T_{ET} \qquad (1.24)$$

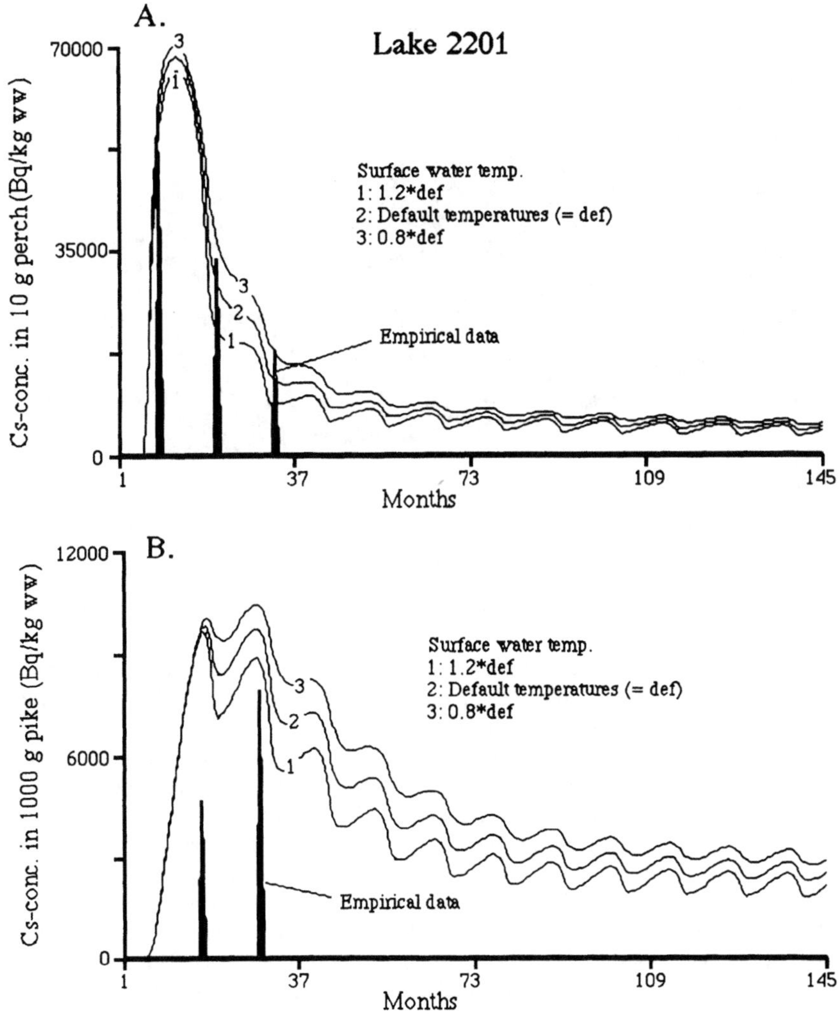

Fig. 1.15. A comparison between empirical data and modelled values for Cs-concentrations in small perch (A) and pike (B) in a scenario where the predicted normal surface water temperatures have been changed by +20% (curve 1) and -20% (curve 3), as compared to the default conditions.

where

M_{ET} = the amount of radiocesium in ET-sediments (Bq);

Vd = the form factor (= 3·Dm/Dmax, where Dm = mean depth and Dmax = max. depth; m); Vd is often called the volume development (see Figure 1.18 or Håkanson, 1981 for further information). It is used as a distribution coefficient in this model to distribute the resuspended amount of radiocesium from ET-areas (i.e., wind/wave-induced advective resuspension) either to the water compartment or to the compartment called accumulation areas (A-areas). If the lake is U-

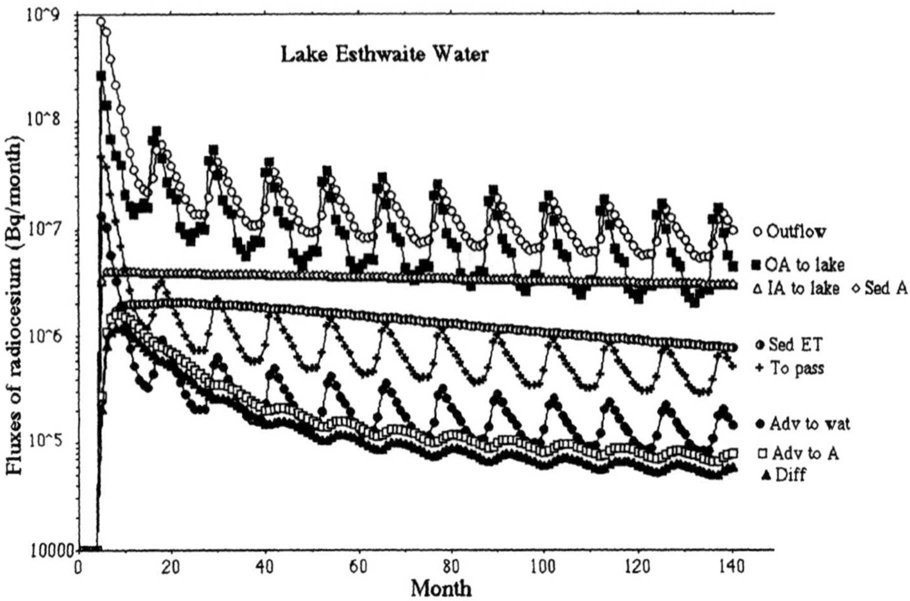

Fig. 1.16. Abiotic radiocesium fluxes in Lake Esthwaite Water.

shaped, Vd is about 3 (i.e., Dmax ≈ Dm) and all resuspended matter will flow to the A-areas. If, on the other hand, the lake is shallow and Vd is small, most resuspended matter will flow to the water compartment.

T_{ET} = age of radiocesium on ET-areas. By definition (see Håkanson and Jansson, 1983), the materials which settle on ET-areas will not stay permanently where they were deposited but will be resuspended by wind/wave activity and/or slope processes. If the age of the material is set to a very long period, e.g., 10 years, these areas will function as accumulation areas; if, on the other hand, the age is set to 1 week or less, they will act as erosion areas. Often (see Håkanson and Peters, 1995), it is assumed that the mean age of these deposits is about 1 year for lakes (and about one month for coastal areas; see Håkanson, 1999). This value is also used as a default value in this model. This means that the halflife, T_{ET}, is given by:

$$T_{ET} = (0.5 \cdot 1 \cdot 12)/0.693 \qquad (1.24)$$

A very important part of this model, or any similar model, is the structure for the partition coefficient, often called Kd (Santschi, and Honeyman, 1991; Riise et al., 1990; Erel and Stolper 1993; Gustafsson and Gschwend, 1997). Traditionally, this is the ratio between the particulate (C'_{part} in Bq/kg dw) and the dissolved (C_{diss} in Bq/1) phases, i.e., $Kd = C'_{part}/C_{diss}$. The total amount (in Bq/l) is equal to ($C_{diss} + C_{part}$), where C_{part} is given in Bq/l. Then Kd is given in l/kg. The amount of suspended matter in the lake water, SPM, is given in mg/l. This means that the dissolved fraction (D_{diss}) can be written as:

$$D_{diss} = 1/(1 + Kd \cdot SPM \cdot 10^{-6}) \qquad (1.25)$$

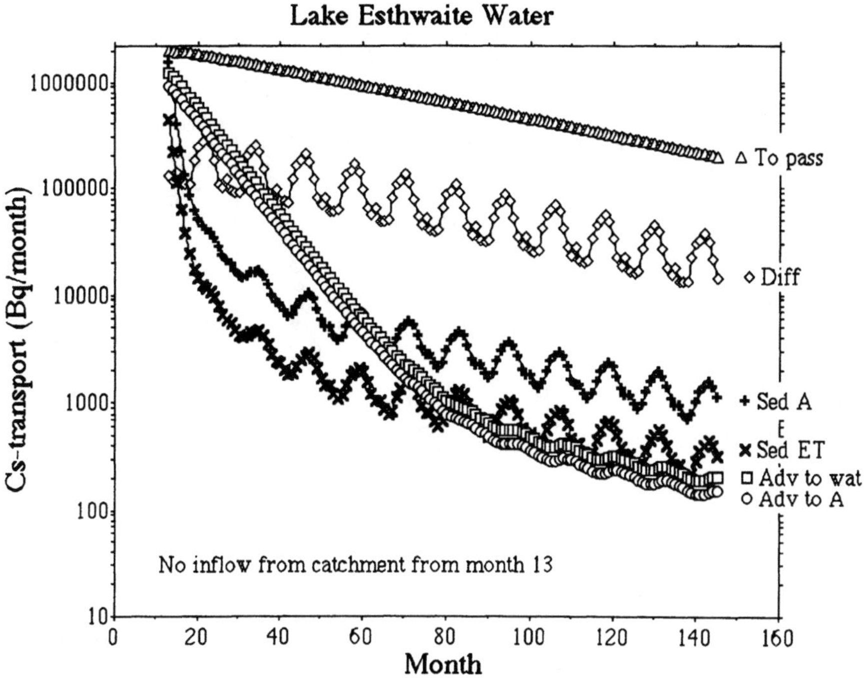

Fig. 1.17. Abiotic radiocesium fluxes in Lake Esthwaite Water when the inflow to the lake was closed in month 13 (= January 1987).

This is a general definition of the dissolved phase based on lake Kd. So, the amount in dissolved form in the lake is equal to C_{diss} (Bq/m^3) multiplied by the total lake volume (m^3).

Values of suspended particulate matter (SPM in mg/l) in the water phase should be empirically measured or predicted from a model (see Table 1.5). The empirical model, which is derived from a broad set of data (see Lindström et al., 1999), gave an r^2 of 0.87. The most important model parameter (explaining 74% of the mean SPM variability among lakes) is the concentration of total phosphorus, indicating that the autochthonous production is the most important process determining the SPM in these lakes. Lake pH (i.e., a collective variable for aggregation and allochthonous and autochthonous processes influencing SPM) and the dynamic ratio (i.e., resuspension; DR = $\sqrt{}$Area/Dm, where Area is in km^2 and mean depth, Dm, is in m) both add significantly to the predictive power of the model. The unexplained residual, 13%, may be accounted for by analytical and sampling variability, and the related uncertainty in the empirical data.

The range of the model parameters for this model is:

Parameter	Unit	Min. value	Max. value
TP	[µg/l]	5	60
pH	[-]	5.10	8.50
DR	[-]	0.07	7.88

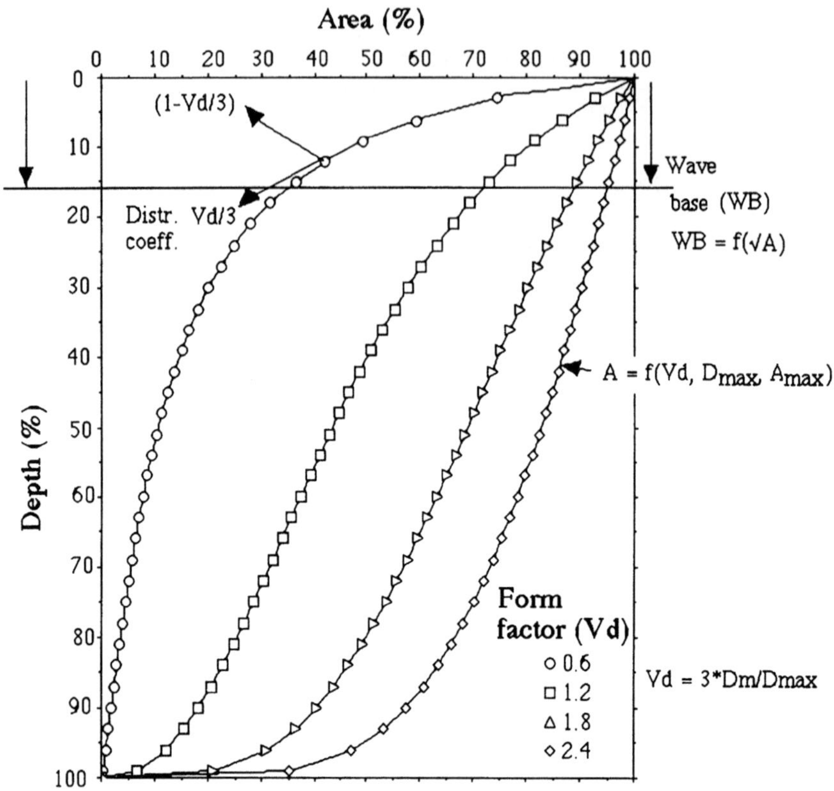

$$Area = A_{max} * ((D_{max} - WB)/(D_{max} + WB * \exp(3 - Vd^{1.5})))^{0.5/Vd}$$

Fig. 1.18. An illustration of how the form factor (= the volume development), Vd, can be used to express the form, here given by the relative hypsographic curve (= the depth-area curve), of lakes. Shallow lakes with a small Vd have relatively large areas above the wave base, where processes of wind/wave-induced resuspension will influence the bottom dynamic conditions. Deep, U-formed lakes generally have smaller areas above the wave base (ET-areas). Modified from Håkanson (1999).

Table 1.5. An empirical approach to predict SPM. Results of the stepwise multiple regression. N = 26 lakes (from Lindström et al., 1999).

Step	r^2	F	Model variable	Model
1	0.74	67	TP	log(SPM) = 1.56·log(TP) - 1.69
2	0.83	11	pH	log(SPM) = 1.47·log(TP) + 0.18·pH - 2.85
3	0.87	6	DR	log(SPM) = 1.148·log(TP) + 0.137·pH + 0.286·log(DR) - 1.985

DR = √Area/Dm; Dm = mean depth.

The default value for lake Kd for radiocesium is often set to about 800,000 l/kg or 0.8 l/mg susp. (from IAEA, 1999). Another approach for radiocesium (see Comans and Hockley, 1992; Smith et al., 1997; Comans et al., 1997) is to express Kd as a function of the "frayed edge site" concentration of the clay mineral illite (which has a unique potential to bind cesium), C_{FES}, the K-concentration of the water, C_K, and the selectivity coefficient between radiocesium and potassium (Kc), where $Kd = Kc \cdot C_{FES}/C_K$. Kc and C_{FES} are often assumed to be constants since it is generally very difficult to access reliable empirical data for these variables for different lakes. This means that the approach may be written as: $Kd = const/C_K$.

On the basis of these arguments, the following approach is meant as a typical "collective" approach, i.e., as a simplification of many complex relationships, but it is also meant to be based on a firm mechanistic foundation. Thus:

$$Kd = 800000/C_K \qquad (1.26)$$

where C_K is given in mg K/ l. 800,000 is the Kd-constant.

1.3.2.2. Lake outflow

The lake outflow is a very important processes: "How open is the exit gate for X from the lake?" This question is handled mathematically by a flux called F_{WO}, i.e., the flux out of the lake (Bq/month).

$$F_{WO} = (R_d + R_W) \cdot M_W \qquad (1.27)$$

where
R_d = the physical decay;
R_W = the lake water retention rate (1/month);
M_W = the amount of radiocesium in the lake water (Bq).

R_W is sometimes set to $1/Tw^1$ in mass-balance calculations. For large, deep lakes with small drainage areas, i.e., lakes with a long theoretical water retention time (Tw), it is evident that thermal and chemical stratifications, hydrological flow patterns and currents influence the retention time of water and contaminants. Then, should one use the entire lake volume, the volume of the epilimnetic water, or the volume of a defined fraction of the epilimnetic water in the calculation of the retention time? From lake eutrophication studies (Vollenweider, 1968), it is well known that better predictions of lake TP-concentrations are obtained if one uses $\sqrt{Tw}$ rather than $KT \cdot Tw$ in mass-balance calculations. In this model, we will define a function (from Håkanson and Peters, 1995) for the exponent in the expression $1/Tw^{exp}$. This exponent (exp) should be about 1 in lakes with a quick water retention (if Tw < 1 months).

- For lakes with very fast water turnover (if Tw < 1 month),

$$R_W = 1.386/Tw \qquad (1.28)$$

- For very small lakes (if Area $< 0.2 \cdot 10^6$ m^2),

$$R_W = 1.386/Tw^{(50/(Tw+50-1) + 0.5)/1.5)))} \qquad (1.29)$$

where 50 is the retention rate constant.

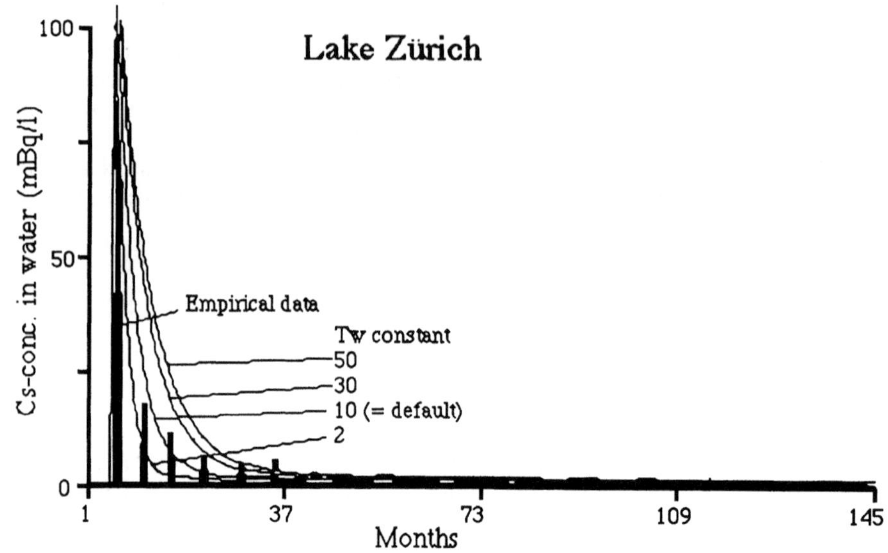

Fig. 1.19. A comparison between empirical data and modelled values for Cs-concentrations in water in Lake Zürich when the retention rate constant has been changed from 2 to 50 (10 is the default value).

• For all other lakes,

$$R_W = 1.386/Tw^{((10/(Tw+10-1) + 0.5)/1.5)))} \quad (1.30)$$

where 10 is the retention rate constant. This approach yields a faster and more adequate retention in lakes with long water retention as compared to 1/Tw. If Tw is 12 months, R_W (eq. 1.30) = 0.28; if Tw = 60 months, $R_W = 0.24$.

The constant 1.386 is, as before, the ratio -ln(0.5)/0.5 = 0.693/0.5, and emanates from the definition of the halflife.

Figure 1.19 illustrates the role that the retention rate constant plays for the retention rate and, hence, for the transport of radiocesium out of the lake and the concentration in the lake. If the retention rate constant is high, the outflow is low and the Cs-concentration in the lake is high, if all else is constant. Figure 1.19 gives model predictions for Lake Zürich, one of the validation lakes. It should be noted that there is a very good correspondence between empirical data and modelled values in Figure 1.19. It should also be noted that this is a validation, not a calibration.

1.3.2.3. Sedimentation on A-areas

We will first consider sedimentation on A-areas (F_{WA}, as defined in eq. 1.16).

$$F_{WA} = M_W \cdot R_{sed} \cdot (1\text{-}ET) \quad (1.31)$$

where

The ET sub-model

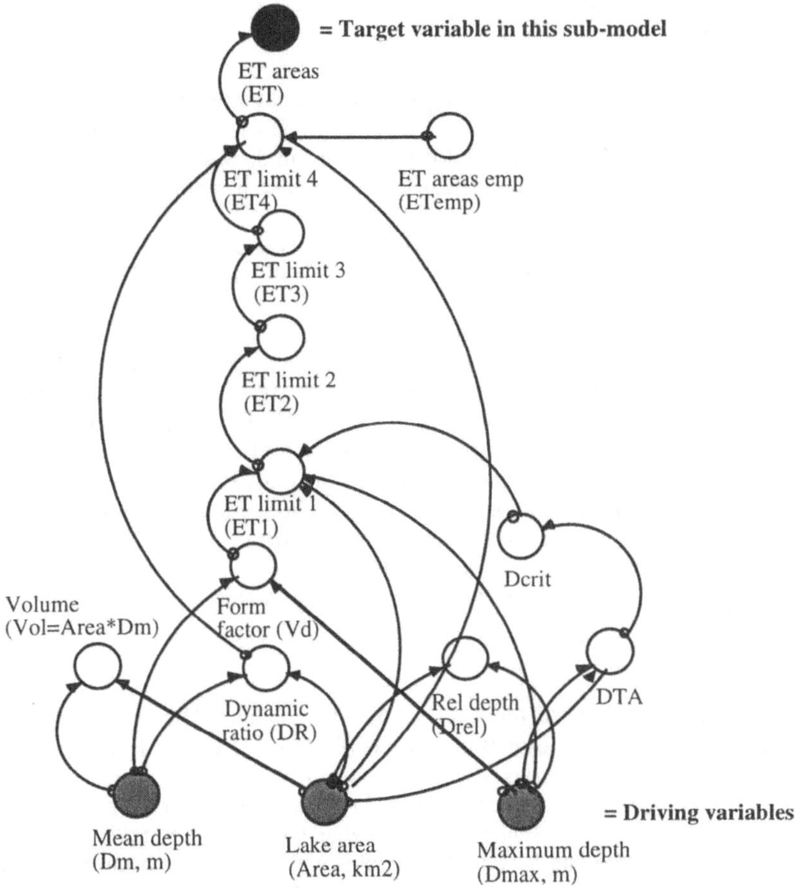

<u>Equations:</u>

DR = √Area/Dm

Drel = (Dmax*√π)/(20*√Area)

If (45.7*√Area)/(21.4+√Area) > Dmax then DTA = Dmax else
DTA = WB = (45.7*√Area)/(21.4+√Area)

If DTA < 0.5 then Dcrit = 0.5 else Dcrit = DTA

Vd = 3*Dm/Dmax

ET1 =
1-(Area*(10^6)*0.01*(100*EXP(-3+Vd^1.5)*(Dmax-Dcrit)/(Dcrit+Dmax*EXP(-3+Vd^1.5)))^(0.5/Vd)*10^(2-1/Vd))/(Area*10^6)

If ET1 > 0.99 then ET2 = 0.99 else ET2 = ET1

If ET2 < 0.15 then ET3 = 0.15 else ET3 = ET2

If ETemp > 0 then ET4 = ETemp else if Area ≥ 1 km2 then ET4 = 0.25*DR*41^(0.061/DR) or if Area < 1 km2 then
ET4 = ET3

Fig. 1.20. The sub-model for predicting ET-areas (see Håkanson and Peters, 1995, for further information).

M_W = the amount of radiocesium in the lake water (Bq);
R_{sed} = the sedimentation rate (1/month);
ET = the fraction of ET-areas, which can, preferably, be determined empirically (see Håkanson and Jansson, 1993, for methods) or predicted from a sub-model (see Figure 1.20).

The following processes influencing internal loading are accounted for in this sub-model for the ET-areas (see Håkanson, 1977; Håkanson and Jansson, 1983):

(1) an energy factor related to the effective fetch and the wave base,

(2) a form factor related to the percentage of the lake bed above the wave base, and

(3) a lake slope factor related to the fact that slope-induced transportation (turbidity currents) may appear on bottoms inclining more than 4-5%.

This information is used to estimate the bottom areas where processes of erosion (E), transport (T) and accumulation (A) prevail. E-bottoms are areas where the cohesive materials that follow Stokes's law are not deposited. Such areas are dominated by coarse deposits. T-bottoms are, by definition, areas where the fine suspended materials are deposited discontinuously. In such areas, one generally finds mixed deposits. Soft A-bottoms appear beneath the wave base, where the fine suspended materials may be continuously deposited. By definition, there should be no net deposition of sediments on E- and T-areas over longer periods of time (1 year or more).

One approach (see Figure 1.20) is used to calculate ET for lakes larger than 1 km^2. This is the equation based on the dynamic ratio (DR), where:

$$ET = 0.25 \cdot DR \cdot 41^{(0.061/DR)} \quad (1.32)$$

Another approach is used for lakes smaller than 1 km^2. This is the approach based on the wave base and the form of the lake, as illustrated in Figure 1.18.

The value used for the ET-areas is used as a distribution coefficient. It regulates the sedimentation of particulate radiocesium either to A-areas or to ET-areas. For many lakes the predictions are rather insensitive to this value. One example of this is given in Figure 1.20 for Lake Esthwaite Water. By definition, ET must vary from 0.15, since there must always be a shallow shore zone where processes of erosion and transport dominate the bottom dynamic conditions, to 1 in large and shallow lakes totally dominated by ET-areas. Figure 1.21 shows that in Lake Esthwaite Water, the predicted Cs-concentrations in water are rather independent of the value selected for ET. The main reason for this is that the internal loading is small in this lake compared to the secondary loading, i.e., the transport from the catchment. Had the tributary transport been small, the value used for ET would have been much more important for the predicted Cs-concentrations in water and fish. It should also be noted that there is a very good correspondence between empirical data and modelled values in Figure 1.20. There has been no tuning of the model to achieve these predictions. Lake Esthwaite Water is a validation lake.

The sedimentation rate, R_{sed}, is given by:

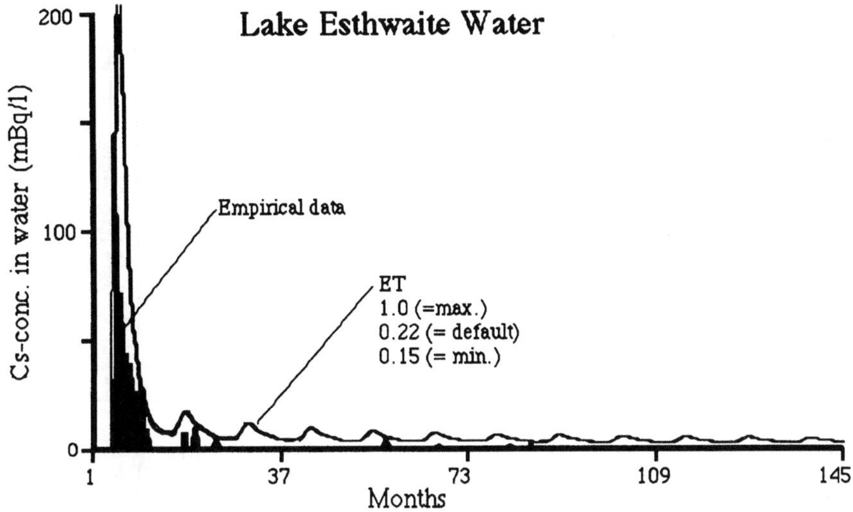

Fig. 1.21. A comparison between empirical data and modelled values for Cs-concentrations in water in Lake Esthwaite Water when the ET-areas have been changed from 0.15, the minimum value in any lake, to 0.22, the actual value in this lake, to 1.0, the maximum value.

$$R_{sed} = ((1-D_{diss})\cdot v_{Cs})/Dm \tag{1.33}$$

where

D_{diss} = the dissolved fraction; this means that $(1-D_{diss})$ is the particulate fraction, which is the only fraction that can settle out in the lake;
v_{Cs} = the default fall velocity for the particulate fraction (m/month);
Dm = the mean depth (m).

Burban et al. (1989, 1990) have demonstrated that changes in turbulence and concentration of suspended particulate matter (SPM) are key regulatory factors for the fall velocity for suspended matter. In this model, the SPM-concentration will be allowed to influence the settling velocity for particulate Cs. This is achieved by means of a dimensionless moderator. The amplitude value is calibrated so that a change in the concentration of SPM by a factor of 10, e.g., from 2 (which is a typical value for oligotrophic lakes) to 20 mg/l (which is typical for eutrophic lakes; see Håkanson and Peters, 1995), will cause a change in the settling velocity by a factor of 3. The borderline value for the moderator is set to 50 mg/l, since it is unlikely that lakes will have a higher suspended amount than that. The default settling velocity for particulate radiocesium is set to 12 m/yr (from Håkanson et al., 1996a). Thus, v_{Cs} is given by:

$$v_{Cs} = (12/12)\cdot Y_{SPM} \tag{1.34}$$

where the dimensionless moderator expressing the influence of SPM on the fall velocity, Y_{SPM}, is given by:

$$Y_{SPM} = (1+0.75\cdot(SPM/50-1)) \tag{1.35}$$

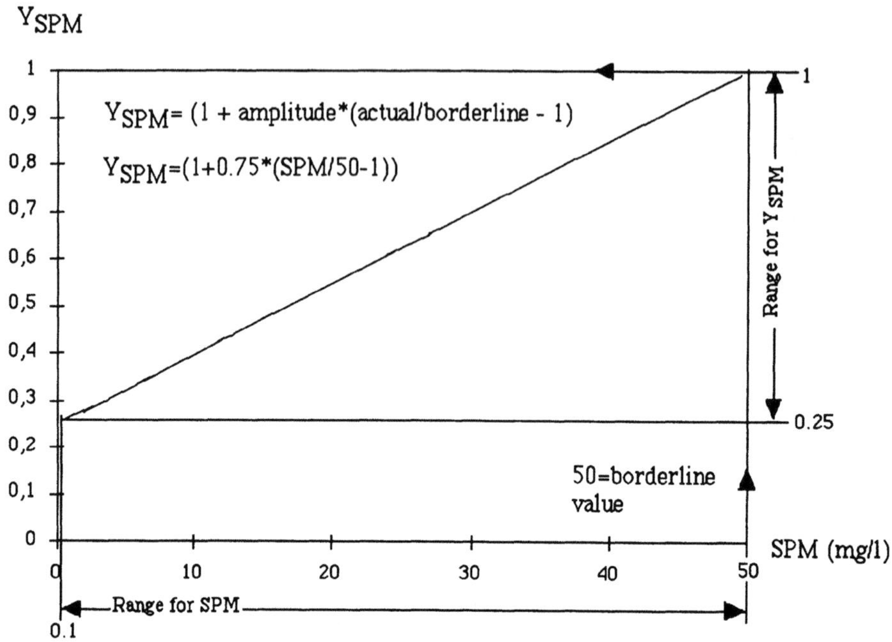

Fig. 1.22. An illustration of the dimensionless borderline moderator for suspended particulate matter, Y_{SPM}.

The calibrated amplitude value is 0.75.

A simple and typical form of a dimensionless moderator is, e.g., the ratio between a mean month value, MM, and a mean annual value, AM. In traditional mass-balance models, one would multiply an amount (kg) by a rate (1/month) to get a flux (i.e., amount·rate = or amount·rate·1). In this model, one multiplies kg·(1/month)·Y (= amount·rate·mod), where Y is a dimensionless moderator quantifying how an environmental variable (like SPM) influences the given flux (e.g., sedimentation of particulate radiocesium). Instead of building a large mechanistic sub-model for how environmental factors influence given rates, this technique uses a simple, general algorithm for the moderator. Empirical data can be used for the calibration of the moderator. The dimensionless moderator defined in this way uses a borderline value, i.e., a realistic maximum value of SPM = 50, to define when the moderator, Y_{SPM}, attains the value of 1. For all SPM-values smaller than the borderline value, Y_{SPM} is smaller than unity (see Figure 1.22). One can also build normal value moderators in such a way that Y_{SPM} is 1 for the "normal" value and higher or lower than 1 for SPM-values higher and lower than the defined normal value (e.g., SPM = 5). The amplitude value regulates the change in Y_{SPM} when the actual SPM-value differs from the borderline value and/or the normal value.

The settling velocity of particulate radiocesium also depends on the amount of resuspended matter. The resuspended particles have already been aggregated; they have also generally been influenced by benthic activities, which will create a "gluing effect" (see Håkanson and Jansson, 1983). It is most probable that such particles have a higher settling velocity than the primary materials. In this model, these resuspended particles settle twice as fast as the primary materials. This is calculated by means of the distribution coefficient (D_{res}), which is illustrated in Figure 1.11, and defined as the ratio between the resuspension (= advection) from ET-areas to water and to A-areas relative to the sedimentation on ET- and A-areas ((AdvW+AdvA)/(SedET+SedA), see Figure 1.11).

The amount of suspended particulate matter (SPM) is important, not so much for the total Cs-concentrations in water, but certainly for the Cs-concentrations in biota. One example of that is given in Figure 1.23, which illustrates how the model predicts Cs-concentrations in water (Figure 1.23A) and in 100 g roach in Lake Hillesjön, Sweden, one of the calibration lakes (see Table 1.3). The test gives Cs-concentrations in water and in roach when the actual SPM-value, 5 mg/l, has been changed by a factor of 5. The higher the SPM-value, the greater the sedimentation of particulate radiocesium, the more cesium in the sediments, the more cesium in benthivores and piscivores, but the less cesium in planktivores and omnivores, like roach. Again, it is worth noting the very good correspondence between empirical data and modelled values.

These results are important and can be illustrated even better by using different values for the default fall velocity (v_{Cs}) in a lake for which there are data on Cs-concentrations in the sediments, piscivorous fish (1000 g pike) and planktivorous fish (30 g perch), namely Lake Flatsjön (figures 1.24 and 1.25). When the sedimentation of particulate cesium increases (either from an increased amount of SPM or by using a larger value for the default fall velocity), the Cs-concentrations in the water decrease (Figure 1.24A), the Cs-concentrations in the active A-sediments increase (Figure 1.24B), the Cs-concentrations in piscivorous pike increase (Figure 1.25A) and the Cs-concentrations in planktivorous perch decrease (Figure 1.25B). All this is logical, and, for this calibration lake, it is worth noting the correspondence between empirical data and modelled values. In these simulations, all model variables have been kept at the generic default values, and only the lake-specific variables have been changed.

Sedimentation of radiocesium on ET-areas, F_{WET} (Bq/month), is, accordingly, given by:

$$F_{WET} = M_W \cdot R_{sed} \cdot ET \qquad (1.36)$$

1.3.2.4. Resuspension from ET-areas to the lake

This is the advective transport of radiocesium from ET-areas, either back to lake water or to A-areas. The form factor, Vd, is used as a distribution coefficient to regulate how much of the resuspended material from ET-areas will go the lake water or to A-areas (see Figure 1.18).

We will first consider resuspension back into lake water, F_{ETW}, i.e., wind/wave-driven advective flux to lake water (Bq/month):

$$F_{ETW} = (M_{ET} \cdot (1\text{-}Vd/3))/T_{ET} \qquad (1.37)$$

where

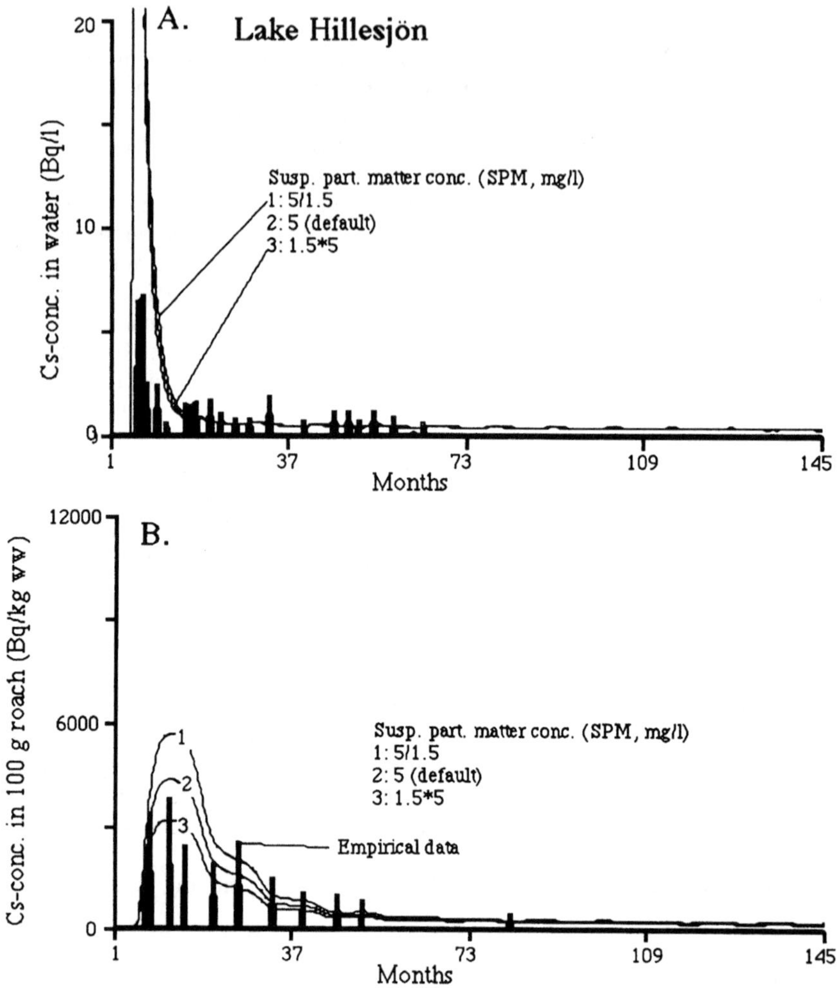

Fig. 1.23. A comparison between empirical data and modelled values for Cs-concentrations in water (A) and in 100 g roach (B) in Lake Hillesjön, Sweden, when the concentration of suspended particulate matter (SPM) is changed by a factor of 5 relative to the actual data.

M_{ET} = the amount of radiocesium in ET-areas (Bq);
Vd = the form factor;
T_{ET} = the age of radiocesium in ET-areas. As mentioned (see Håkanson and Peters, 1995), it is often assumed that the mean age of these deposits is about 1 year for lakes. Note that the calculation time, dt, is given in months, not years. T_{ET} was given by eq. 1.24.

Advective transport (= resuspension) from ET-areas to A-areas, F_{ETA} (Bq per month), is given by:

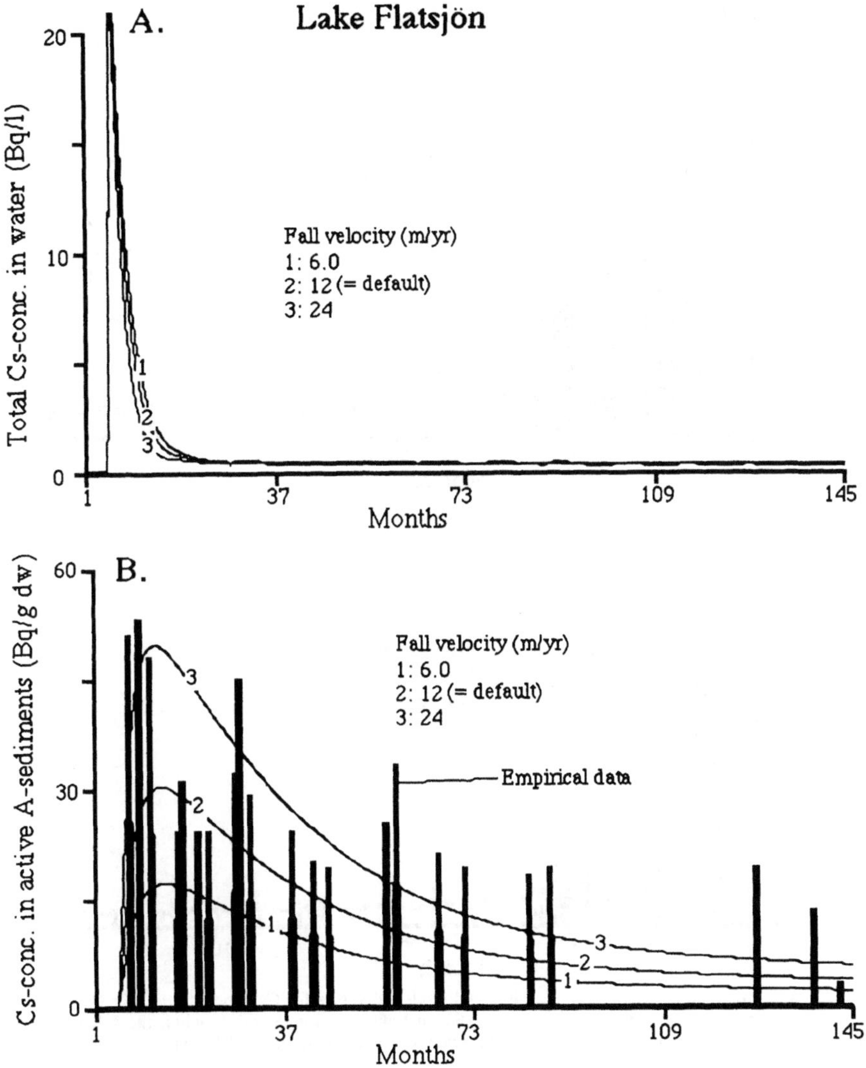

Fig. 1.24. A comparison between empirical data and modelled values for total Cs-concentrations in water (A) and in active A-sediments (B) in Lake Flatsjön, Sweden, when the fall velocity of particulate cesium (v_{Cs}) is changed by a factor of 2 relative to the default value (12 m/yr).

$$F_{ETA} = (M_{ET} \cdot Vd/3)/T_{ET} \quad (1.38)$$

This equation completes the set of equations regulating the fluxes of radiocesium to and from the compartment lake water.

Next, we will consider the two sedimentological compartments, i.e., radiocesium associated with A- and ET-areas.

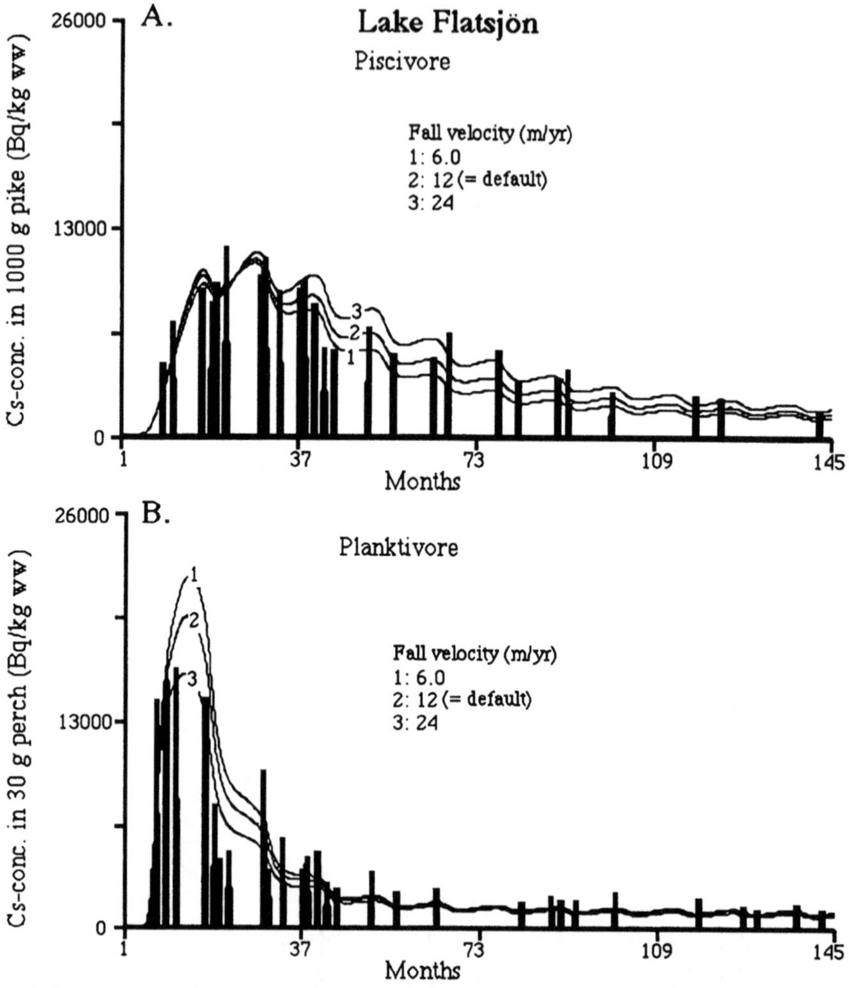

Fig. 1.25. A comparison between empirical data and modelled values for Cs-concentrations in 1000 g pike, a piscivore (A), and in 30 g perch, a planktivore (B), in Lake Flatsjön, Sweden, when the fall velocity of particulate cesium (v_{Cs}) is changed by a factor of 2 relative to the default value (12 m/yr).

1.3.2.5. Active A-sediments

The following differential equation gives all fluxes to and from the compartment, i.e., active A-sediments:

$$M_A(t) = M_A(t - dt) + (F_{WA} + F_{ETA} - F_{WD} - F_{APS})\cdot dt \qquad (1.39)$$

All these fluxes, except for radiocesium transport from surficial, active sediments to deeper, geological (= passive) sediments, F_{APS}, have been treated.

The age of the active A-sediments is needed to calculate the retention rate of the substance in this compartment, and, hence, the flux out of the compartment to passive (geological) sediments (the retention rate is set to 1/age). The age is generally calculated as the ratio between the depth of the active sediments (in m) and the deposition of materials on A-areas (in m/month). Assuming a default value for the depth of the active layer as 5 cm and a default sedimentation rate of 4 mm/yr for A-areas (see Håkanson and Peters, 1995), we have a default value for the age of the active sediments, T_A, of 150 months, i.e., the retention rate, $1/T_A$, is given by:

$$1/T_A = 0.693/(0.5 \cdot 150) \tag{1.40}$$

The biological mixing from the benthos, gas ebullition, etc., will create transport of old, previously deposited sediment into the compartment of the active A-sediments. This will influence the age of the active sediments. It is well known from lake sedimentological studies (see Håkanson and Jansson, 1983) that the benthos can eat the bottom sediments many times over (up to 7 times). When new materials are deposited on the sediments, these will be compacted and the water content will increase with sediment depth (see Håkanson and Jansson, 1983). This means that the actual age of the active sediments is generally significantly older than that indicated by the simple ratio between sedimentation (cm/yr) and the thickness of the active sediment layer (cm). The correction factor for bioturbation is related to the gross sedimentation (see Håkanson and Jansson, 1983). If the deposition is very high, the bottom-living animals are less likely to be able to create complete mixing.

The following factors are known to be important for sedimentation and, hence, also for the age of the active sediments in lakes (Håkanson and Jansson, 1983; Hilton, 1985; Evans and Håkanson, 1992):

1. A <u>load</u> factor describing the natural load of allochthonous materials and the anthropogenic load (from industries, urban areas and agricultural activities) on a given lake.

2. A <u>pre-trapping</u> factor describing the fact that lakes "downstream" receive less allochthonous materials than lakes "upstream", which act as sediment traps for suspended particles/aggregates.

3. A <u>production</u> factor describing the autochthonous particle production, i.e., the internal bioproduction.

4. A <u>deposition</u> factor describing the capacity of a given lake to act as a sediment trap - the larger the lake volume and the longer the water retention time, the higher the entrapment capacity if all other factors are held constant.

5. A <u>resuspension</u> factor linked to the morphometry of the lake, the topographical characteristics of the lake surroundings and the wind climate.

Evans and Håkanson (1992) demonstrated that the average, recent sedimentation in lakes could be predicted from a model based on (1) mean pH as a water chemical indicator (of very many chemical, biological and physical processes in the lake and its catchment area), (2) maximum depth as a link to lake morphometry and the potential for resuspension, and (3) lake area as a measure of lake size and lake productivity.

Table 1.6. Data from 25 Swedish lakes used to develop the empirical model for gross sedimentation in sediment traps (SedB). ADA = area of drainage area in km^2, Dmax = max. depth in m, Area = lake area in km^2, Dm = mean depth in m, Drel = relative depth, dim. less, Tw = theoretical lake water retention time, yr, pH = characteristic lake pH, TP = characteristic lake total-P concentration, μg/l, col = characteristic lake colour, mg Pt/l, SedB, the target y-variable, gross sedimentation in bottom-placed sediment traps in μg dw/cm^2·day.

Lake	Name	ADA	Dmax	Area	Dm	Drel	Tw	pH	col	totP	SedB
701	Skärsjön	0.92	4.6	0.36	2.4	0.68	1.36	6.5	14	6.6	114
703	Hacksjön	4.04	2.4	0.26	1.4	0.42	0.38	6.2	58	13.3	251
705	Skogsrydss.	12.42	2.7	0.33	1.7	0.42	0.18	5.7	173	34.7	673
706	Kinnen	28.08	16.2	1.44	5.1	1.20	1.16	6.2	63	12.2	110
1804	Ölsjön	44.16	25.2	0.74	8.7	2.60	0.47	5.7	68	16.0	29
1808	Stora Ånsjön	20.15	3.2	0.25	2.2	0.57	0.07	6.0	81	8.0	134
1814	Grytsjön	1.63	2.9	0.22	1.3	0.55	0.50	4.8	258	32.0	50
1818	Bergtjärn	0.56	13.5	0.04	5.3	5.98	0.15	5.5	107	11.0	28
1819	Trehörningen	0.39	18.4	0.06	6.2	6.66	1.86	6.0	82	10.0	39
1820	Sarvtjärnen	0.36	25.0	0.09	8.1	7.39	3.77	6.1	81	12.0	34
2110	Ecklingen	13.30	19.2	0.44	4.7	2.57	0.48	6.6	112	11.8	175
2117	Tansen	14.40	13.0	0.58	4.2	1.51	0.53	5.6	130	10.6	44
2119	Bältbosjön	1.00	6.8	0.15	2.7	1.56	1.20	5.9	79	8.9	73
2120	Långsjön	3.30	6.2	0.16	2.1	1.37	0.30	6.2	133	8.5	118
2121	Blacksåstjärn	1.90	6.8	0.12	2.4	1.74	0.47	6.4	115	11.6	65
2122	Bottentjärn	3.60	4.9	0.16	2.8	1.09	0.38	5.8	110	10.0	75
2201	Selasjön	20.05	16.3	0.25	8.3	2.89	0.33	5.2	122	9.5	28
2206	Rävsjön	25.05	21.3	0.59	6.4	2.46	0.48	5.6	128	11.9	56
2212	Stor-Habborn	17.60	19.5	0.28	4.9	3.27	0.24	6.0	79	8.7	34
2213	V. Lövsjön	4.36	19.2	0.47	5.1	2.48	1.60	6.4	38	7.1	58
2214	Lill-Selasjön	38.48	9.2	0.07	2.9	3.08	0.02	5.1	115	10.1	28
2215	Herrbodtjärn	8.34	12.6	0.24	4.5	2.28	0.44	6.2	70	9.5	56
2216	Lill-Bandsjön	6.04	11.2	0.16	4.0	2.48	0.37	6.5	35	11.1	101
2217	Hamstasjön	11.57	7.8	0.18	3.8	1.63	0.21	6.6	46	22.9	431
2218	V. Långedsjön	16.83	4.3	0.25	2.1	0.76	0.10	6.5	69	13.8	116
Min.		0.36	2.4	0.04	1.3	1.06	0.02	4.8	14	6.6	28
Max.		44.16	25.2	1.44	8.7	1.86	3.77	6.6	258	34.7	673
Mean		11.94	11.7	0.32	4.1	1.85	0.68	6.0	95	12.9	117
SD		11.88	7.2	0.29	2.1	1.18	0.79	0.5	49	6.8	143

This section presents a new empirical model for sedimentation in lakes. It is partly based on the same empirical data-set used by Evans and Håkanson (1992). One strict requirement of this model is that it must only be driven by variables already included in the overall lake model for radiocesium.

The data used to derive the empirical model for gross sedimentation incorporated in this model for radiocesium are given in Table 1.6. The model for gross sedimentation in lakes is presented in Table 1.7. The method used to derive this empirical model, including the stepwise multiple regression analysis, is given in Håkanson and Peters (1995). This model is based on data on gross sedimentation from sediment traps placed in the deepest parts of the lakes (SedB). Thus, the equation provides maximum values of gross sedimentation. It is well known (see Håkanson and Jansson, 1983) that sedimentation increases from about zero at the limit between areas of transport and areas of accumulation to maximum values at the deepest part of the basin, a phenomenon often referred to as "sediment focusing". In this approach there is a correction factor applied to the model-predicted values based on this knowledge. The predicted values for gross sedimentation are assumed to be correct for a U-shaped basin, with a form factor of 3, and too low for a V-shaped basin with a form factor smaller than 3. thus, the ratio Vd/3 is used as a correction factor.

Table 1.7. An empirical model to predict gross sedimentation in bottom-placed sediment traps (SedB in µg dw/cm^2·day) based on data from 25 Swedish lakes (data from Håkanson and Peters. 1995; F > 4).

Step	r^2	Model variable	Model
1	0.45	Drel	log(SedB)=2.05 - 0.7·log(Drel)
2	0.66	pH	log(SedB)=-0.745-0.679·log(Drel)+0.354·pH
3	0.79	TP	log(SedB)=-1.56-0.53·log(Drel)+0.4445·pH+0.8524·log(TP)

This means that the model for gross sedimentation is:

$$GS = (Vd/3)\cdot SedB \qquad (1.41)$$

and SedB = the gross sedimentation in sediment traps, as determined from the empirical model:

$$SedB = 10^{(-1.56-0.53\cdot \log(Drel)+0.4445\cdot pH+0.8524\cdot \log(TP))} \qquad (1.42)$$

This empirical model gave an r^2-value of 0.79 for these 25 Swedish lakes. The model applies if the model variables are within the following ranges:

	Drel	**pH**	**total-P (µg/l)**
Min.	1.06	4.8	6.6
Max.	1.86	6.6	35

The values for pH and TP should be the mean, characteristic lake values, e.g., the mean annual values.

Thus, gross sedimentation increases with:

1. Relative depth (Drel = $Dmax \cdot \sqrt{\pi}/(20 \cdot \sqrt{Area})$; see Håkanson, 1981), which is directly related to internal loading (resuspension) and the form of the lake. The shallower the lake, the smaller the Drel-value, the larger the resuspension, and the larger the gross sedimentation.

2. Lake pH, which influences the aggregation of suspended particles, and many other lake processes, e.g., the higher lake pH, the more productive the lake, and the higher the gross sedimentation.

3. Lake total-P concentration, which is directly related to lake production.

If gross sedimentation (GS) is larger than 2000 µg dw/cm^2·day, it is assumed that the oxygen consumption is so high that the oxygen concentration is smaller than 2 mg/l and bioturbation is halted because the meio- and macrobenthos die.

The bioturbation factor (BF) is given by:

if GS > 2000 then BF = 1 else

$$BF = (4-0.02 \cdot (GS/100-1)) \quad (1.43)$$

where 4 is the bioturbation and compaction constant, 0.02 is the amplitude value and 100 is the norm-value for GS.

If the actual GS increases from the norm-value of 100 to 1000, the BF-value changes from 4 to 3.18. This means that the mean age of the sediments is likely to be about 20% lower allowing for these changes in bioturbation and compaction.

These factors, i.e., gross sedimentation, bioturbation and compaction, will influence the age of the active A-sediments (see Figure 1.26). This means that F_{APS}, is given by:

$$F_{APS} = M_A \cdot (1/(BF \cdot T_A) + R_d) \quad (1.44)$$

where R_d is the physical decay rate for radiocesium.

The concentration of radiocesium in active A-sediment can be calculated in a standard fashion. This requires data on the volume of the active A-sediments, sediment water content and sediment bulk density, as given in Figure 1.26. In this model, one can use a water content of 70% ww and an organic content (= loss on ignition) of 7% dw as default values to describe the physical character of the active A-sediments (0-5 cm).

1.3.2.6. Areas of erosion and transport

The following differential equation gives all fluxes to and from the compartment, i.e., ET-areas:

$$M_{ET}(t) = M_{ET}(t - dt) + (F_{WET} - F_{ETW} - F_{ETA} - F_{ETD}) \cdot dt \quad (1.45)$$

All these fluxes, except for radiocesium transport from ET-sediments by physical decay, F_{ETD}, have been considered previously, and F_{ETD}, is given by: $F_{ETD} = M_{ET} \cdot R_d$.

Thus, all the abiotic fluxes are explained, and we can turn to the biological part of the model.

Sub-model for gross sedimentation and sediment age

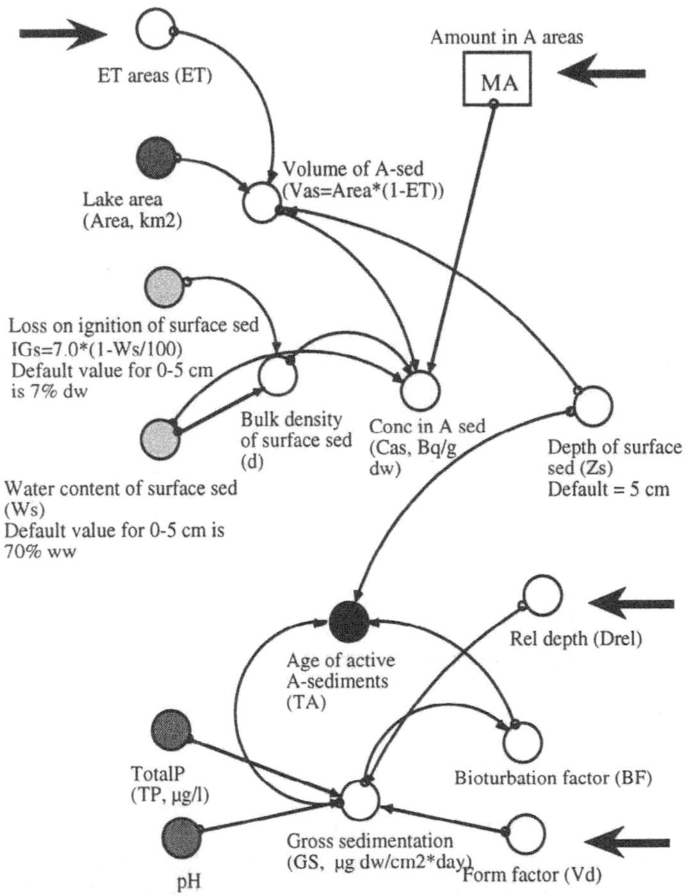

Equations:
Cas=MA/(Vas*d*(1-Ws/100))
TA=BF*Zs*((1-Ws/100)*Ds*12)/(GS)
GS=(Vd/3)*10^(-1.56-0.53*log(Drel)+0.4445*pH+0.8524*log(TP))

Dep of materials cm per yr, Dep=GS(1/(1-Ws/100))*(1/d)

Bulk density of surface sed,
Ds=100*2.6/(100+(Ws+IGs)*(2.6-1))

If GS > 2000 then BF = 1 else BF = (4-0.02*(GS/100-1))

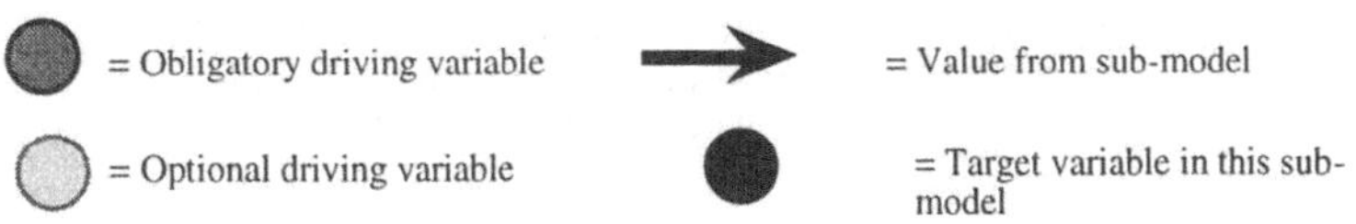

Fig. 1.26. The sub-model for gross sedimentation and sediment age.

Fish sub-model

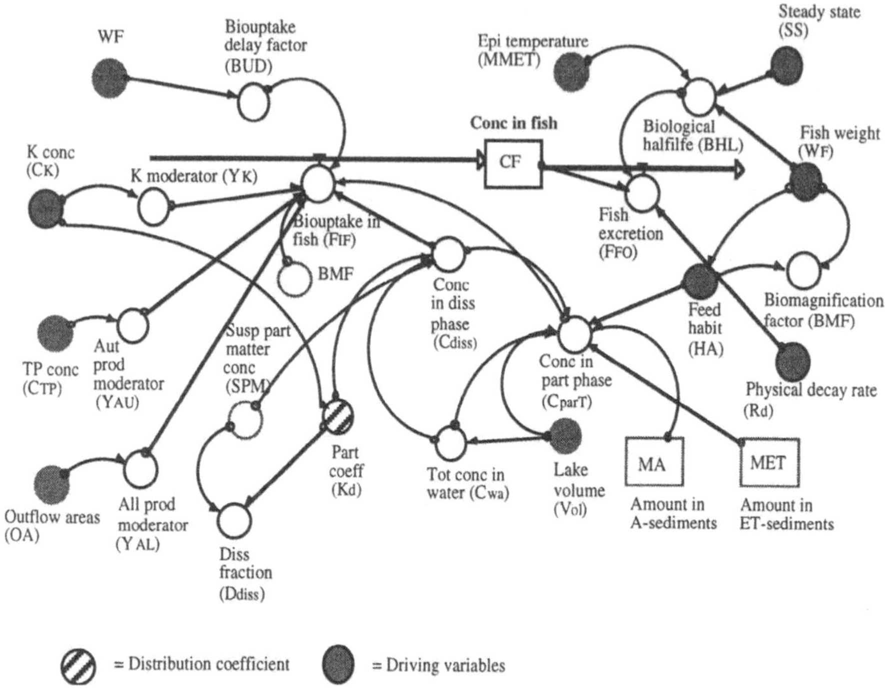

Fig. 1.27. An outline of the biotic part of the model for radiocesium.

1.3.3. BIOTIC PROCESSES

1.3.3.1. General introduction

There is one sub-model for biota, for fish, in this model.

Technically, the biotic processes are treated separately from the biotic fluxes and the two pathways are linked by means of a bioconcentration factor (Figure 1.1B). This is a simplification with many advantages in predictive modelling:

1. One can then predict the target concentrations in biota directly without data on biomasses, which are generally extremely hard to access. The basic aim here is to predict the concentration of X in biota. This model uses compartments of one mass unit: ^{137}Cs (= X) is transported from one mass unit of lake water to one mass unit of biota. This means that the model automatically yields concentrations. Most mass-balance models use the lake or a given part of the lake as the unit of size and calculate the concentration of a substance in water as the ratio between the total amount (g or kg) of material in the total volume (l or m^3). Such models can be extended to the populations of pike or perch by treating the fish as summed volumes (e.g., V_1 or V_2 in l = dm^3). If the density = d (kg/l = g/cm^3), and if X_1 and X_2 are the amounts of contaminant (in g), then the concentrations of X in pike and perch are $X_1/V_1 \cdot d$ and $X_2/V_2 \cdot d$, respectively, in the two species.

2. One can relatively easily test (calibrate and validate) models of this type, since concentrations in biota are easily measured.

This model structure generally gives better predictive success than traditional models using connected or parallel biotic compartments (like phytoplankton, zooplankton, benthos, prey fish and predator fish models of the type illustrated in Figure 1.1A). The uptake of X from water to biota is still modelled dynamically.

1.3.3.2. The fish model

In the fish model, the bioconcentration factor is modified by factors known to influence biouptake of ^{137}Cs:

- The potassium concentration in lake water.
- The amount in dissolved and particulate phases of radiocesium.
- The feeding habits of the fish.
- The weight of the fish.
- The trophic characteristics of the lake.
- The water temperature.

These factors are accounted for in this model, which is illustrated in Figure 1.27.

Approaches like this are most relevant in cases when the biotic fluxes of X are much smaller than the abiotic fluxes. This is the case for practically all metals and radionuclides in most lakes (see Håkanson, 1999).

The differential equation for the concentration (i.e., C_F in Bq ^{137}Cs per kg wet weight of fish) is given by:

$$C_F(t) = C_F(t - dt) + (F_{IF} - F_{FO})\cdot dt \quad (1.46)$$

where
F_{IF} = biouptake (Bq/month·kg);
F_{FO} = excretion (Bq/month·kg).

The biouptake is given by:

$$F_{IF} = Y_{AL}\cdot Y_{AU}\cdot Y_K\cdot BMF\cdot SMTH(C_{diss}+C_{parT},BUD,C_{diss}+C_{parT}) \quad (1.47)$$

where
Y_{AL} = a dimensionless moderator for allochthonous production;
Y_{AU} = a dimensionless moderator for autochthonous production;
Y_K = a dimensionless potassium moderator;
BMF = a biomagnification factor;
SMTH = a smoothing function based on the concentration of radiocesium in dissolved phase in lake water, C_{dissT}, the total amount in particulate phase in lake water and sediments, C_{par}, and a biouptake delay factor (BUD).

These factors affecting the biouptake will now be examined.

The moderator for allochthonous production, Y_{AL}, as well as the moderator for autochthonous production, Y_{AU}, are meant to account for the principle of "biological dilution" (see Håkanson and Peters, 1995) whereby a given load of a radionuclide or a metal into a lake will be distributed over a larger biomass in a lake with higher bioproduction than in a lake with a lower bioproduction. The bioproduction in, "brown-water" lakes in particular (see Jonsson, 1997) is not, however, a matter of just the supply of total phosphorus, handled by Y_{AU} in this approach. Nitrogen can often be the limiting nutrient in such lakes, and Y_{AL} is meant to account for this in a simple manner. It would, perhaps, have been logical to account for this by including more model variables, like total-N concentration, lake colour, etc. However, it is a well established fact (see Håkanson and Peters, 1995) that "brown-water" lakes generally have relatively large areas of wetland (mires, bogs, etc.), which supply the lakes with coloured substances (humus). The moderator for allochthonous production, Y_{AL}, is based on the percentage of outflow areas, since this information is included in this model for other reasons (in the catchment area sub-model). This means that allochthonous production is accounted for without adding more driving variables - an important requirement of this model. By incorporating this moderator for allochthonous production, the model becomes more general. Empirical calibrations have, however, demonstrated that it is not important to account for this effect in lakes barely influenced by humic materials. That is, if the outflow area, OA, is smaller than the calibrated limit of 0.1 (10% of the catchment area), the allochthonous influences on the bioconcentration factor for radiocesium are likely to be very small. If OA > 0.1, it becomes increasingly important to account for this principle. Then, this moderator accounts for allochthonous production by the ratio 0.1/OA. It should be noted that this moderator will cause an decrease in biouptake by a factor of 2 if OA is 0.2 (values of OA > 0.3 would be very rare; see Håkanson and Peters, 1995). So, for Y_A:

- if OA < 0.1, then $Y_{AL} = 1$
- if OA ≥ 0.1, then $Y_{AL} = 0.1/OA$ (1.48)

Empirical calibrations (the calibration lakes are presented in Table 1.3) for radiocesium have demonstrated that it is important also to account for autochthonous phytoplankton production, especially for low-productive lakes; if lake total-P concentration is lower than about 14 μg/l, then the following applies:

- if TP > 14 (μg/l), then $Y_{AU} = 8$
- if TP ≤ 14, then $Y_{AU} = (22\text{-TP})$ (1.49)

This moderator accounts for autochthonous production in a linear way. If, e.g., the total-P concentration is very low, say 4 μg/l, which is typical for ultra-oligotrophic lakes, eq. 1.49 sets the moderator to 16. It the TP-concentration is higher than 14, the moderator is 8 and the expected concentration of radiocesium in fish becomes lower.

Figure 1.28 illustrates how different lake TP-concentrations, and hence different levels of lake bioproduction, influence the predicted Cs-concentration in fish, here 100 g roach in Lake Flatsjön (one of the calibration lakes in Table 1.3). The higher the bioproduction, the more pronounced the biological dilution and the lower the Cs-concentrations in fish. For this calibration lake, it is worth noting the good correspondence between empirical data and modelled values. In these simulations, all model variables have been kept at the generic default values, and only the lake-specific variables have been changed.

The potassium moderator, Y_K, expresses the influence of lake K-concentrations on the biouptake for fish. Note that the same moderator may be used for all types of biota. The critical limit between sufficiency and deficiency is generally (see Fernandez et al., 1997) set at $C_K = 0.1$ mM = 3.91 mg K/l. In natural lakes, one can expect C_K to

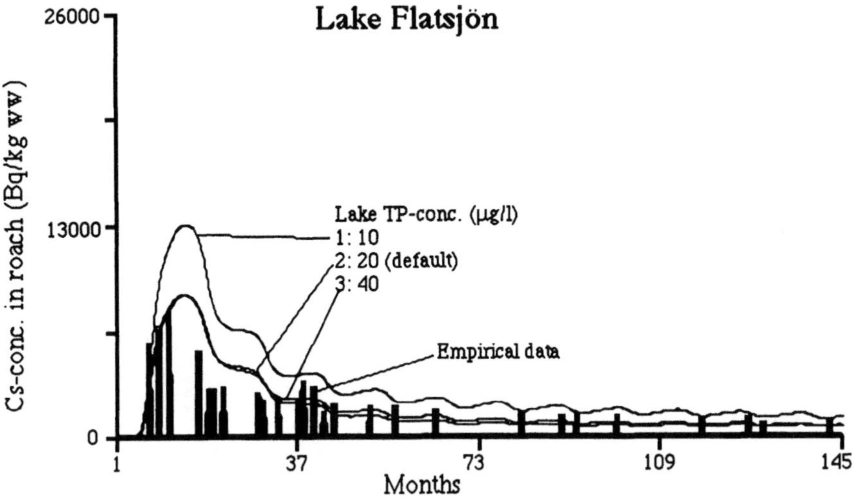

Fig. 1.28. A comparison between empirical data and modelled values for Cs-concentrations in 100 g roach in Lake Flatsjön, Sweden, when the lake TP-concentration is changed by a factor of 2 relative to the actual data (20 μg/l in this lake).

vary from about 0.1 to 150 mg/l (see Desmet, 1997). The dimensionless moderator is based on physiological mechanistic models for the bioconcentration factor for both sufficiency ($C_K > 0.1$ mM), using the Nernst equation, and deficiency ($C_K < 0.1$ mM), using the Michaelis-Menten model (see Figure 1.29).

The entire realistic range for C_K was tested in the derivation of the K-moderator (see Figure 1.30). The curve marked CF/CF_{norm}, i.e., the ratio between the actual calculated concentration factor (CF) and the reference, or norm CF-value ($CF_{norm} = 40$; the ratio is used to get a dimensionless moderator), is the theoretical dimensionless moderator. This curve is calculated from the Michaelis-Menten equation for deficiency and the Nernst equation for sufficiency according to the presuppositions given in Figure 1.29. The halflife for Cs in cells is set to 0.55 days, the default temperature to 20°C, the default Cs-concentration in water to 0.001 μM, the maximum uptake rate in cells to 46,560 μmol/gdw·day, the affinity constant of the active transport system for Cs (Ks_{Cs}) to 27.5 μM, the affinity constant of the active transport system for potassium to 23.2 μM, and the diffusion potential for K to 105 mV. There is also an empirical constant in the Nernst equation for radiocesium. The constant (0.73) comes from a linear fitting of CF in *Riccia fluitans*, subjected to potassium sufficiency, as a function of the external potassium concentration. This parameter modifies the capacity of cells to accumulate Cs vs K, i.e., it is a selectivity coefficient. It agrees roughly with the relative permeability of Cs over K (0.6) obtained from controlled physiological experiments.

Figure 1.30 shows the smooth transition from sufficiency to deficiency, which indicates that the model is well calibrated. Figure 1.30 also illustrates that the calculated curve (CF/CF_{norm}) can, in fact, be described very well by a dimensionless moderator, Y_K, expressed as:

$$Y_K = 300/(4.5+C_K^{0.75}) \tag{1.50}$$

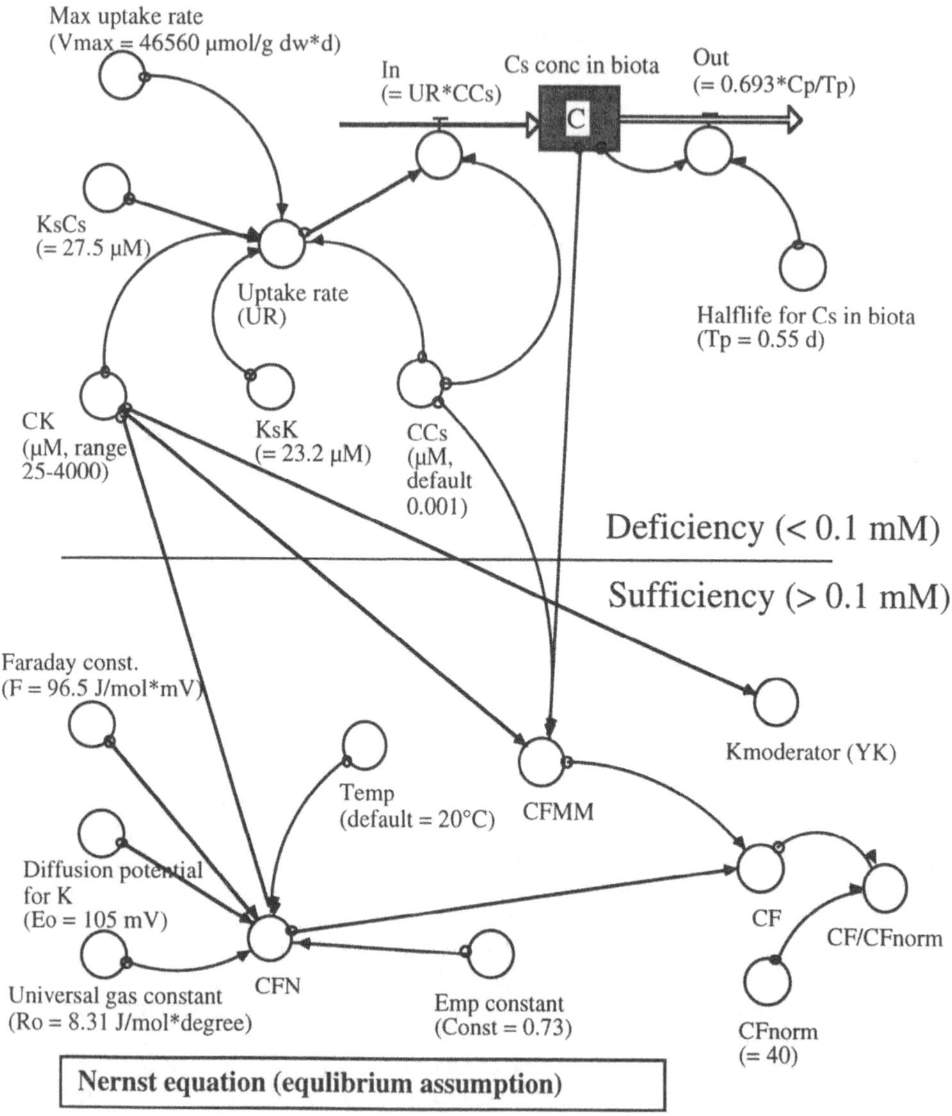

Equations:
UR=Vmax/(KsCs*(1+(CK/KsK))+CCs)
If CK ≥ 100 then CF = CFN =
1/EXP(Const*ln(CK*0.001)+(-Eo*F/(R*(Temp+273))))
IF CK < 100 then CF = CFMM = Cp/CCs

Fig. 1.29. The Michaelis-Menten equation and the Nernst equation.

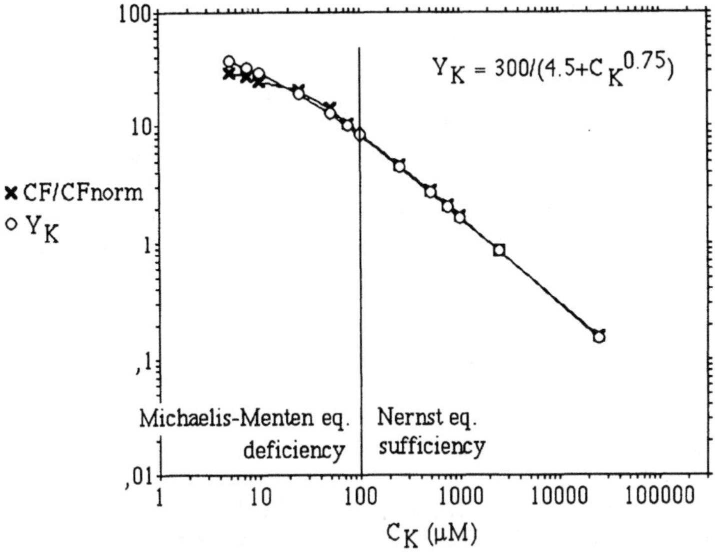

Fig. 1.30. A comparison between model-predicted values for the CF/CF_{norm}-ratio using the Michaelis-Menten equation and the Nernst equation, and the new dimensionless moderator for potassium (Y_K).

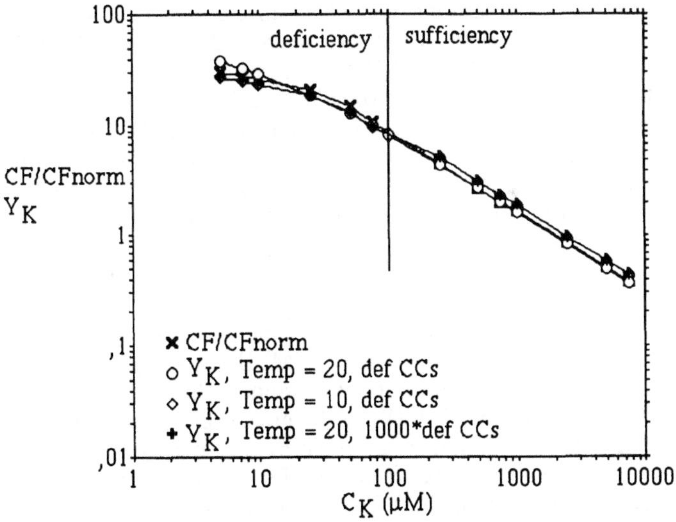

Fig. 1.31. A comparison between model-predicted values for the CF/CF_{norm}-ratio using the Michaelis-Menten equation and the Nernst equation, and the new dimensionless moderator for potassium (Y_K) for two temperatures (10 and 20°C) and two Cs-concentrations (default = 0.001 μM and 1 μM).

This is meant to be a general moderator applicable to plants and animals. It does not depend on temperature or on the Cs-activity in the water, as illustrated in Figure 1.31, which gives predicted CF/CF_{norm}-ratios for the default conditions and for a temperature of 10 °C and a Cs-concentration 1000 times higher than the default value. Note that in this model C_K must be given in µeq/l (= (mg/l)/0.0391).

Figure 1.32 illustrates how the K-moderator actually works. If the Y_K-value is set higher than the default value, as given by eq. 1.50, the biouptake increases and the Cs-concentrations in biota, here 500 g pike in Lake Flatsjön (one of the calibration lakes in Table 1.3), increases. Figure 1.25A gave the same prediction for 1000 g pike. It should be noted that there is a very good correspondence between empirical data and modelled values and that the predictions for 1000 g pike agree better with the empirical data than the predictions for 500 g pike. In these simulations, all model variables have been kept at the generic default values, and only the lake-specific variables have been changed.

The biomagnification factor (BMF) is determined accordingly (partly from Rowan and Rasmussen, 1994, 1995):

if the fish weight (WF) is < 10 g ww, then BMF = 25 for all types of fish; if WF > 10 g ww, then:

250/WF	for planktivores
2	for benthivores
3	for omnivores and
4	for piscivores.

The biomagnification factor accounts for differences in biouptake/biomagnification from water to fish related to feeding habits when all else is constant. Fish of the same weight but with different feeding habits display different biouptake. Piscivores are, e.g., likely to take up more radiocesium than planktivores. If everything else is constant except for the feeding behaviour expressed by these trophic categories, the factors 2, 3 and 4 should reflect the most likely differences in concentrations. The BHL-value for planktivores seems to depend very much on the weight of the fish.

The biouptake delay factor (BUD) accounts for the fact that large/old animals high up in the nutrient chain do not respond as quickly to changes in concentrations in water as small/young planktivores (see e.g., Håkanson, 1991). This delay is expressed as a function of the weight of the animal (WF in g ww). In this model where the calculation time, dt, is set to one month, the delay factor is set to 1 for small fish (weighing less than 100 g ww). For larger fish, BUD is given by:

$$BUD = WF/100 \tag{1.51}$$

BUD is used as an averaging time within the framework of a smoothing function (see Figure 1.13). In this model we have: SMTH(C_{diss}+C_{parT},BUD,C_{diss}+C_{parT}), where C_{diss} and C_{parT} are the concentrations of radiocesium in dissolved and particulate phases. The greater the weight of the fish (WF), the greater the value of BUD, and the more pronounced the delay in the uptake, as given by the smoothing function. Figure 1.33 illustrates how the biouptake delay factor (BUD) actually works for whitefish in Lake Iso Valkjärvi, Finland, one of the validation lakes (see Table 1.4). If the reference weight used in eq. 1.51 is changed from the default value of 100 g ww to 50 and 25 g, the curves in Figure 1.33 give the calculated consequences for the concentration in 150 g (planktivore) whitefish. The smaller the reference fish weight, the smaller the biouptake delay. It should also be noted that also in this validation there is a very good correspondence between empirical data and modelled values.

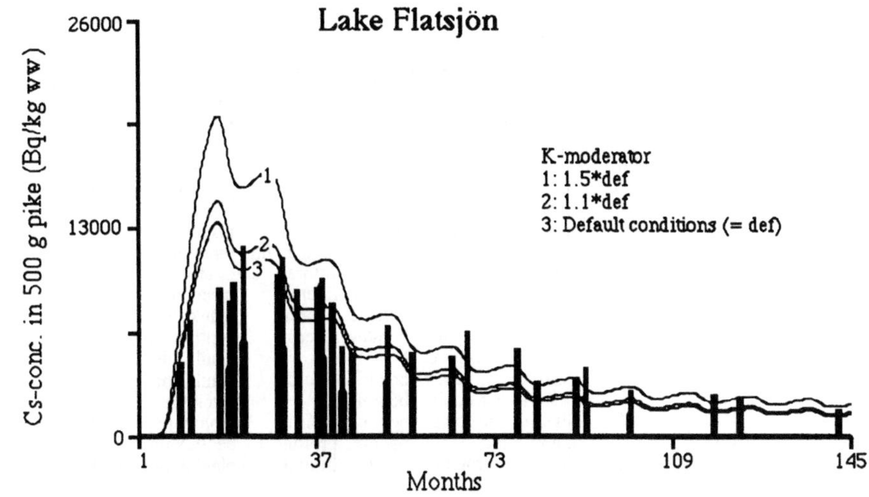

Fig. 1.32. A comparison between empirical data and modelled values for Cs-concentrations in 500 g pike in Lake Flatsjön, Sweden, when the K-moderator is changed from the default value (def, as given by eq. 1.50) to 1.1·def and 1.5·def.

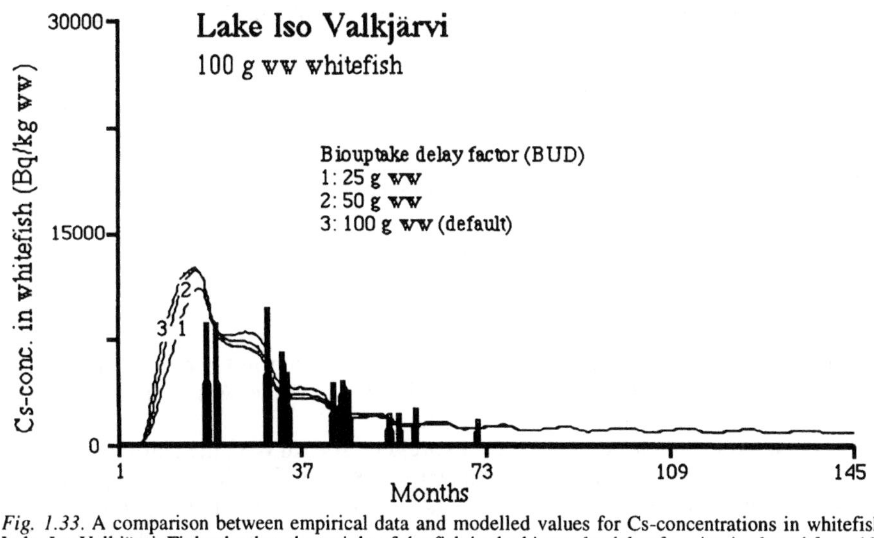

Fig. 1.33. A comparison between empirical data and modelled values for Cs-concentrations in whitefish in Lake Iso Valkjärvi, Finland, when the weight of the fish in the biouptake delay function is altered from 100 g to 50 and 25 g.

So, the weight of the fish is important. Table 1.8 gives a useful compilation of information (from Brittain, 1998) regarding common European species of fish, typical weight ranges and feeding habits. This information can be used to identify feeding habits and typical weights for all species of fish discussed in this book, and many more.

Table 1.8. Compilation of major European fish species in rivers. lakes and reservoirs (modified from Brittain, 1998). CM = consumed by man. I, II and II = stages.

Species	Typical weight range, kg	Target weight kg	Trophic range	Habitat	Food habit	Significance
Sturgeon	20-50	30	meso- to eutrophic	benthic. riverine	piscivore	CM, caviar
Brown trout. I	0.05-0.1	0.1	oligo- to mestotr.	benthic. littoral	planktivore	CM, early
Brown trout. II	0.1-0.3	0.2	oligo- to mestotr.	benthic. littoral	benthivore	CM
Brown trout. III	0.5-1.5	1	oligo- to mestotr.	benthic. littoral	piscivore	CM
Arctic char	0.01-0.2	0.1	oligo- to mestotr.	pelagic	planktivore	CM
Whitefish	0.1-1	0.3	oligo- to mestotr.	pelagic	planktivore	CM
Smelt	0.01-0.05	0.01	oligo- to eutrophic	pelagic	planktivore	prey
Pike	0.5-3	1	oligo- to mestotr.	benthic. littoral	piscivore	CM
Roach	0.05-0.2	0.1	oligo- to hypertr.	benthic/pelagic. litt.	omnivore	prey
Minnow	0.001-0.01	0.01	oligo- to mesotr.	benthic. littoral	omnivore	prey species
Asp	0.5-3	1	meso- to eutrophic	pelagic. riverine	piscivore	CM
Nase	0.1-1	0.3	meso- to eutrophic	pelagic. riverine	planktivore	CM
Barbel	0.5-3	1	meso- to eutrophic	benthic. riverine	benthivore	CM, angling
Bream	0.5-2	1	meso- to eutrophic	benthic. littoral	benthi-/detrivore	CM
Carp	0.5-3	1	meso- to eutrophic	benthic. littoral	benthi-/detrivore	CM
Wels	2-20	5	meso- to eutrophic	benthic	piscivore	CM
Eel	0.1-1	0.5	oligo- to eutrophic	benthic. riverine	omnivore	CM
Burbot	0.1-1	0.5	oligotrophic	benthic. profundal	benthivore	CM
Perch. I stage	0.01-0.1	0.1	oligo- to eutrophic	benthic. littoral	planktivore	CM, early
Perch. II	0.1-0.3	0.2	oligo- to eutrophic	benthic. littoral	benthi-/omni.	CM
Perch. III	0.3-0.6	0.5	oligo- to eutrophic	benthic. littoral	piscivore	CM
Pike-perch	0.5-3	1	oligo- to eutrophic	pelagic	piscivore	CM
Ruffe	0.005-0.02	0.01	oligo- to eutrophic	benthic. littoral	benthivore	prey species

The concentration of radiocesium in the dissolved phase, C_{diss}, is calculated from (see also eq. 1.25):

$$C_{diss} = C_{wa}/(1+Kd \cdot SPM \cdot 10^{-6}) \quad (1.52)$$

where
C_{wa} = the total concentration of radiocesium in lake water (Bq/l).

The total concentration of all particulate phases, C_{parT}, is given by:

$$C_{parT} = HA \cdot ((Cs_{wa}-C_{diss})+0.001 \cdot (M_A+M_{ET})/Vol) \quad (1.53)$$

Hence, C_{parT} is the total concentration of radiocesium in (1) particulate phases suspended in the water plus (2) the amount of ^{137}Cs in active A-sediments and ET-sediments.

The default value used in the basic expression for Kd (i.e., the constant 800,000 in eq. 1.26) and the reliability of the characteristic K-concentration greatly influences the Cs-concentration in biota. Figure 1.34A gives model predictions for Cs-concentration in 200 g perch and Figure 1.34B gives calculated values for total Cs-concentration in water for Lake IJsselmeer, The Netherlands, when the K-concentration is changed from 7 (the actual value) to 14, 3.5 and 1.75 mg K/l. This influences both the distribution coefficient, lake Kd, and hence how much of the cesium in the lake is bound to particles, that can settle out in the lake, and the K-moderator, which regulates biouptake and, hence, Cs-concentrations in fish. There is a very interesting

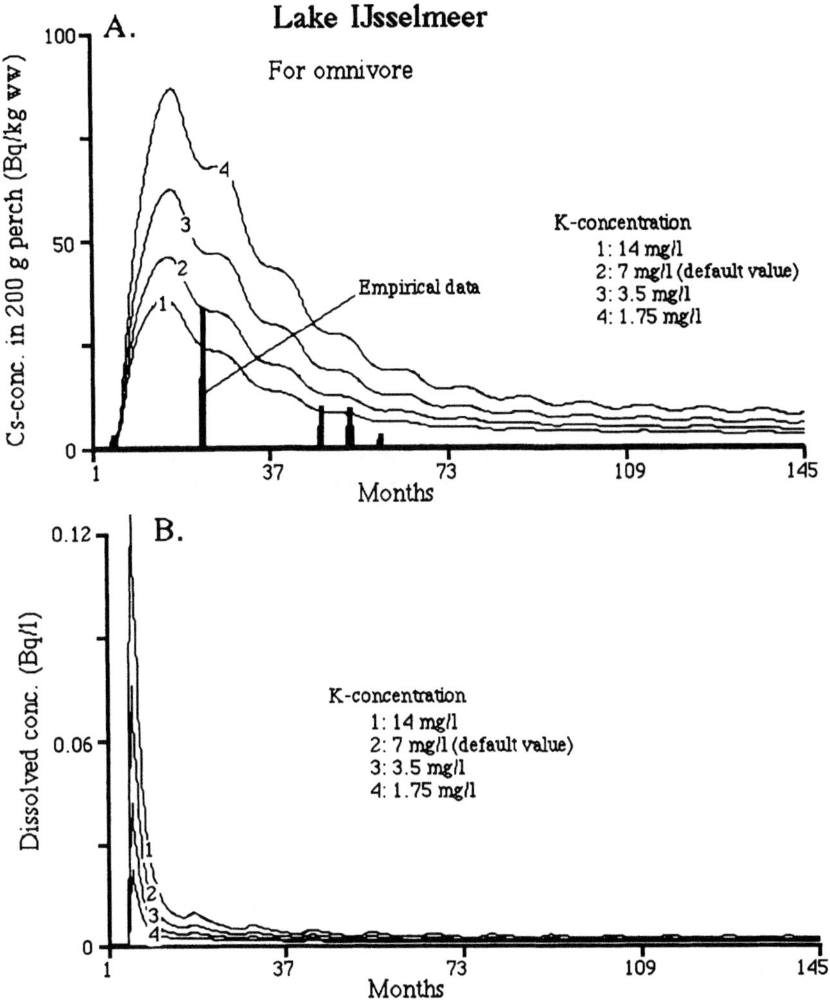

Fig. 1.34. A comparison between empirical data and modelled values for Cs-concentrations in 200 g perch, an omnivore (A), and modelled predicted values for total Cs-concentrations in lake water (B) in Lake IJsselmeer, The Netherlands (one of the validation lakes in Table 1.4), when the K-concentration in water is changed from 7 (actual value) to 14, 3.5 and 1.75 mg/l.

balance involved in these two equations. Figure 1.34A shows that for 200 g perch, an omnivore, the lower K-concentration implies lower Cs-concentrations in fish, but higher Cs-concentrations in water (Figure 1.34B). This means that more radiocesium is deposited in the sediments at higher values of K in the water. This is reflected in the higher Cs-concentrations in the omnivore. On the other hand, the lower concentrations in the water (Figure 1.34B) cause lower Cs-concentrations in planktivorous fish, as exemplified by the 50 g smelt in Figure 1.35A. This shows the very interesting, important

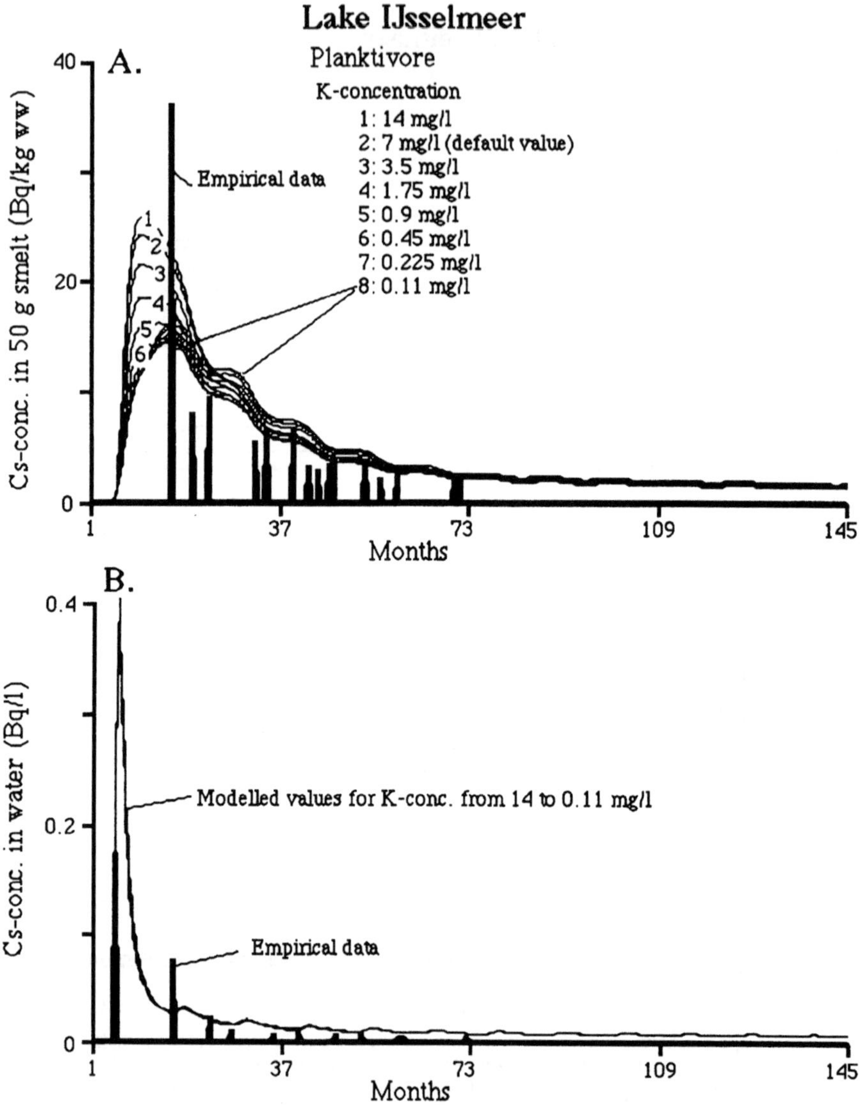

Fig. 1.35. A comparison between empirical data and modelled values for Cs-concentrations in 50 g smelt, a planktivore (A) and total Cs-concentrations in lake water (B) in Lake IJsselmeer, The Netherlands, when the K-concentration in water is changed from 14 to 0.11 mg/l.

but complex role that potassium plays relative to cesium in lake water: it influences both fundamental abiotic and biotic processes.

Lake IJsselmeer is a validation lake (Table 1.4) and it is worth noting the good correspondence between empirical data and modelled values for Cs-concentrations in perch (Figure 1.34A), smelt (1.35A) and water (1.35B).

HA means feeding habit. The values used for HA in this model are meant to reflect the influence on the bioconcentration factor when all factors, except exposure to all types of particulate phases of ^{137}Cs, are held constant. This factor should reflect the importance of radiocesium fluxes from sediments and suspended particulate matter. This means that the factor should be highest for benthivores and lowest for planktivores, and in-between for other groups of fish. In this model, we use the following default values for HA.

Benthivore: HA = 1/2.
Piscivore HA = 1/8.
Omnivore HA = 1/20.
Planktivore HA = 1/200.

The concentration (in Bq/l) of radiocesium, which can be taken up by biota living in ET- and active A-sediments (Csed), is calculated as:

$$C_{sed} = 0.001 \cdot (M_A + M_{ET})/Vol \qquad (1.54)$$

This is the second term in eq. 1.53. The first term is C_{par}.

The excretion, F_{FO} (in Bq/kg ww·month), from the fish is given by:

$$F_{FO} = C_F \cdot (BHL + R_d) \qquad (1.55)$$

where
C_F = the concentration in fish (Bq/kg ww);
BHL = the biological halflife of radiocesium in fish (1/month);
R_d = the physical decay rate (1/month).

This expression for the biological halflife (BHL) comes from Rowan and Rasmussen (1995).

• If WF < 5 g ww, then

$$BHL = 30 \cdot e^{(-6.583-0.111 \cdot \ln(WF \cdot 5)+0.093 \cdot MMET+0.326 \cdot SS)} \qquad (1.56)$$

• If WF ≥ 5 g ww, then

$$BHL = 30 \cdot e^{(-6.583-0.111 \cdot \ln(WF)+0.093 \cdot MMET+0.326 \cdot SS)} \qquad (1.57)$$

That is, the BHL-value depends on the weight of the fish (WF), the epilimnetic water temperature (MMET in °C) and whether steady-state conditions are imminent or not. If the model predictions are based on steady-state assumptions, then SS = 0, if not, then SS = 1. Note that normally one CANNOT assume steady-state.

Figure 1.36A illustrates how the fish weight influences the Cs-concentrations in trout in Lake Øvre Heimdalsvatn, Norway, one of the validation lakes (see Table 1.4). Trout is a planktivore, and the value selected for the weight of the fish is of profound importance for the predictions, i.e., whether predictions agree with the empirical data. From Figure 1.36A, one can see that almost perfect predictions are obtained for the actual fish weight of 200 g in this lake. Figure 1.36B gives a similar sensitivity analysis

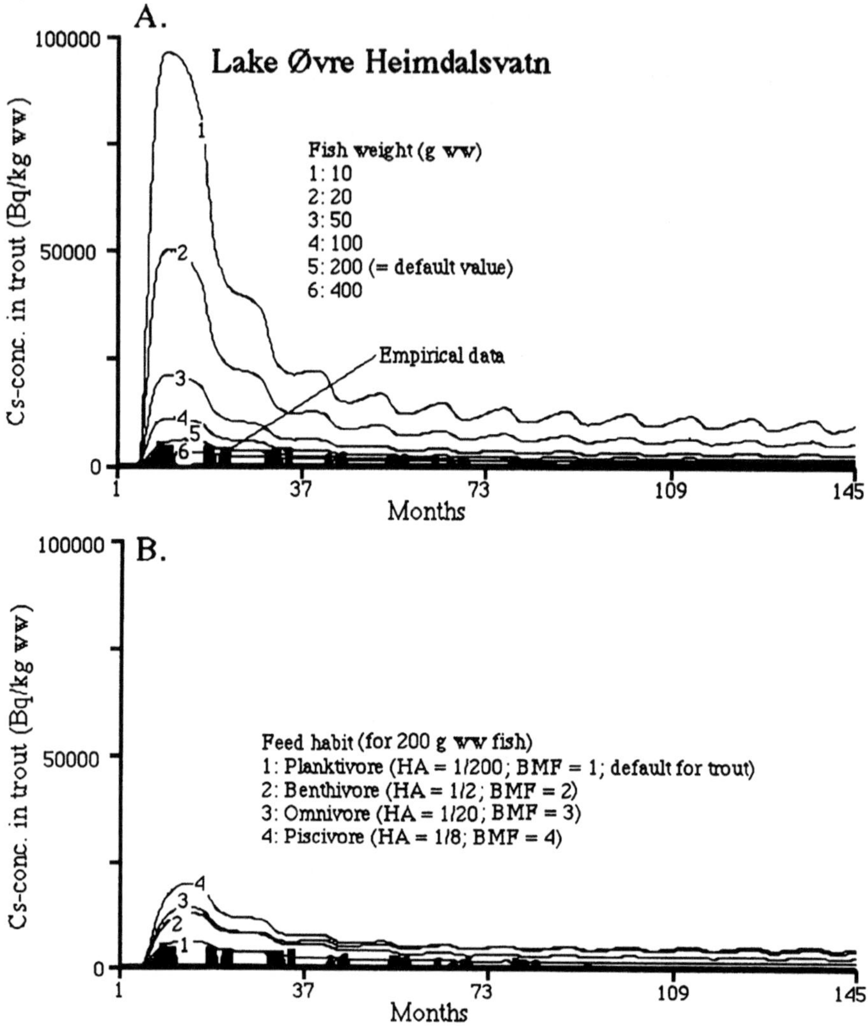

Fig. 1.36. A comparison between empirical data and modelled values for Cs-concentrations in trout (a planktivore) of different weights (A) and for Cs-concentrations in 200 g fish with different feed habits (B) in Lake Øvre Heimdalsvatn, Norway.

when the feeding habit, HA, has been changed. Curve 1 gives predictions for a planktivore, curve 2 for a benthivore, curve 3 for a omnivore and curve 4 for a piscivore (all weighting 200 g ww). There are significant differences depending on the selected feeding habit. The highest predicted Cs-concentrations, when all else is constant, appear for piscivores, the next highest for benthivores, followed by omnnivores and planktivores.

Again, it is worth noting the good correspondence between empirical data and modelled values for these validations.

1.3.4. CRITICAL MODEL TESTS

1.3.4.1. Set-up of the tests

The aim of the following section is to present critical model tests according to the methods discussed by Håkanson and Peters (1995). This model has been calibrated using data for the VAMP lakes listed in Table 1.3. They cover a wide range in terms of size (from 0.16 to 537 km^2), fallout (3 to 130 kBq/m^2) and other lake characteristics.

Sensitivity analysis involves the study, by modelling and simulation, of how an alteration of one rate or variable in a model influences a given prediction, while everything else is kept constant. This type of analysis plays a dominant role in ecosystem modelling (see Hinton, 1993; Hamby, 1995; IAEA, 1999). The previous sections have given many simple sensitivity tests. However, sensitivity analysis is a wider concept and it usually involves, at least, two further steps. The first step in the sensitivity analysis is often to multiply a given x-variable with a uniform factor, like 0.1 and 10, while all else is constant, and thereby study the effects this will have on the target y-variable.

The next step in a sensitivity analysis is often to repeat this type of calculation for all interesting model variables and to use not uniform factors but realistic uncertainty factors for all the x-variables. The CV-values given will be used for this purpose in the following sensitivity and uncertainty tests. It is evident that it is NOT realistic to apply the same uncertainty to all model variables, such as a factor of 10 as in many of the previous examples. There are major differences among model variables in this respect (Table 1.2).

Two main approaches to uncertainty analysis exist - analytical methods (Cox and Baybutt, 1981; Beck and Van Straten, 1983; Worley, 1987) and statistical methods like Monte Carlo techniques (Tiwari and Hobbie, 1976; Rose et al., 1989; IAEA, 1989). In this section, we will only discuss Monte Carlo simulations.

Uncertainty tests using Monte Carlo techniques may be done in several ways, using uniform CV-values or, more realistically, using characteristic CV-values (e.g., from Table 1.2). For predictive empirical or dynamic models based on several uncertain model variables (rates, etc.), the uncertainty in the prediction of the target variable depends on such uncertainties. The cumulative uncertainty from many uncertain x-variables may be calculated by Monte Carlo simulations, and that is the focus of this section.

Monte Carlo simulations are used to forecast the entire range of likely observations in a given situation; it can also give confidence limits to describe the likelihood of a given event. Uncertainty analysis is the same as conducting sensitivity analysis for all given model variables at the same time. A typical uncertainty analysis is carried out in two steps. First, all the model variables are included with defined uncertainties and the resulting uncertainty for the target variable calculated. Then, the model variables are omitted from the analysis one at a time. This is the way in which these tests have been done.

Thus, the following tests are meant to identify critical parts of the model, which regulate the reliability and uncertainty of the model predictions.

The calibrations have yielded suggestions for all the model variables, which are compiled in Figure 1.37, the panel of driving variables. This panel is divided into two

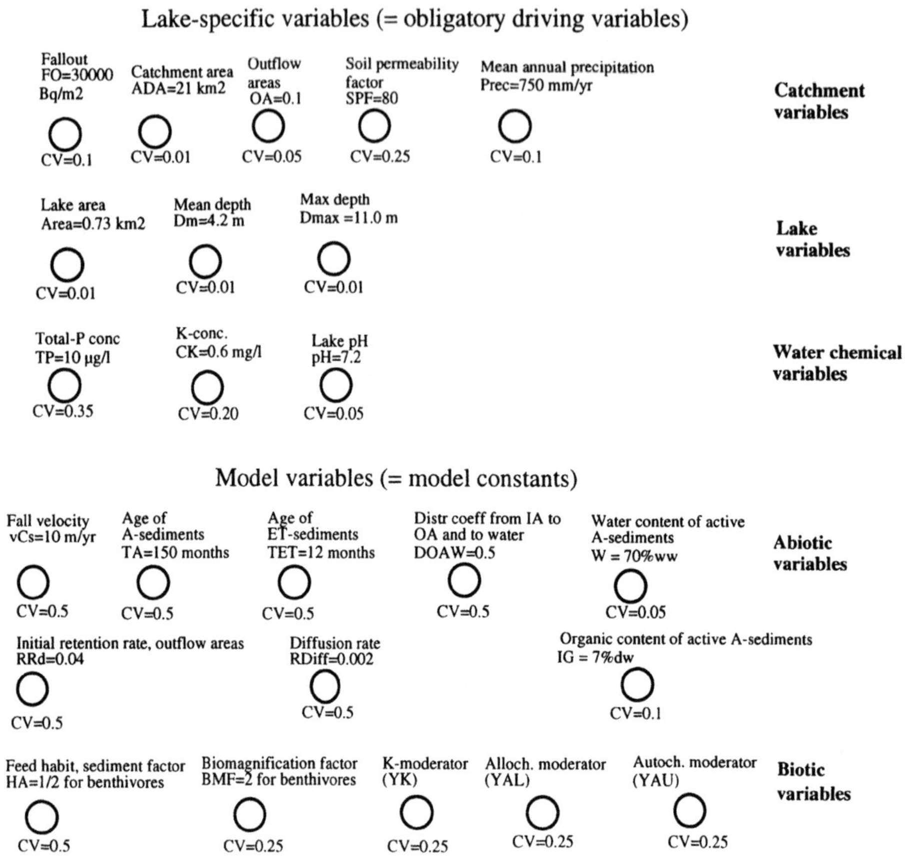

Fig. 1.37. The panel of driving variables for the radiocesium model using lake-specific data for Lake Siggefora, Sweden. The panel also gives characteristic CV-values. Note that information about fish species and the mean population weight of the given fish species must also be given before a simulation can be performed.

main parts - the lake-specific variables, which must be available for every lake (these are the obligatory driving variables), and the model variables (the model constants, which should NOT be changed for different lakes but be regarded as generic model values). The lake-specific variables are divided into:

- catchment variables; fallout, catchment area, outflow areas and soil type (or soil permeability factor), mean annual precipitation, latitude, altitude and continentality;
- lake variables; lake area, mean depth, maximum depth and theoretical lake water retention time;
- water chemical variables; TP-concentration, K-concentration and pH;

The actual data for one of the validation lakes, Lake Siggefora, Sweden, are also given in Figure 1.37, as well as characteristic CV-values (see Table 1.2) for all the model variables, which will be used later in the sensitivity and uncertainty analyses.

The model variables are categorised into:

• abiotic variables: fall velocity (v_{Cs}), age of A-sediments (T_A), age of ET-sediments (T_{ET}), distribution coefficient between inflow and outflow areas (D_{OAW}), the initial retention rate of radiocesium in outflow areas (R_{Rd}), the diffusion rate (R_{Diff}), water content of active (0-5 cm) A-sediments and organic content of active A-sediments;

• biotic variables: feed habit (sediment factor, HA), biomagnification factor (BMF), K-moderator (Y_K), autochthonous moderator (Y_{AU}) and allochthonous moderator (Y_{AL}).

Before a simulation can be done, one must also decide which fish species is to be predicted and the mean population weight of the selected fish species (see Table 1.8).

1.3.4.2. Sensitivity and uncertainty analyses

There are two targets for the following tests: Cs-concentrations in lake water and in fish. Generally in dynamic models, like this radiocesium model, it is more difficult to predict concentrations of toxic substances in biological compartments than in abiotic compartments (see Håkanson and Peters, 1995). So, one can expect the sensitivity and uncertainty tests to demonstrate a smaller CV for the predicted Cs-concentrations in lake water compared to fish. These tests have focused on the model variables, i.e., on the most uncertain parts of the model, these being the rates and distribution coefficients which have the highest CV-values (as given in Figure 1.37). The following variables have been included in the tests.

1. The weight of the fish (Weight). There is always some uncertainty concerning which mean population weight should be used. Generally, however, empirical data on the fish weight would exist. Such empirical data will give a mean weight and a CV for fish weight. This particular uncertainty evidently CANNOT influence the predictions for the Cs-concentrations in water, but it may be important for the predicted Cs-concentrations in fish. The idea with these tests is to see how important this uncertainty is compared to other uncertainties. In the following tests, we will use a CV of 0.2 for fish weight. This means that if the weight of the fish is set to 100 g (ww), the fish weight is likely to be (with a 95% certainty) between 60 and 140 g (corresponding to 2·SD).

2. The biological halflife (BHL). The approach used in this model has been tested in several contexts (see Rowan and Rasmussen 1995; Håkanson et al., 1996a, b; IAEA, 1998). It is fairly well known that BHL depends on the weight of the fish and lake water temperature. The CV is set to 0.1.

3. The biomagnification factor (BMF). The approach used in this model emanate from rather extensive empirical tests (see Rowan and Rasmussen 1994, 1995) and the CV-value for this factor has been set to 0.25, as opposed to 0.5 for most rates in the model, which have not been tested in the same manner. This uncertainty only affects the Cs-concentration in fish.

4. The feeding habit (or sediment factor, HA) reflecting the exposure from all particulate phases (benthic pathways, etc.). This factor is large for benthivores (1/2), smaller for piscivores (1/8) and omnivores (1/20) and very small for planktivores (1/200). The actual values for the different exposures to sediments and particulate matter are, however, uncertain, and HA has been given a CV of 0.5. Also this uncertainty can only affect the Cs-concentration in fish.

5. The mean monthly epilimnetic temperature (MMET). Even if very reliable data for the mean monthly surface water temperatures are available from long measurement series, such data may not be correct for future lake temperatures. Information on mean monthly surface water temperatures is generally available from hydrological and meteorological tables or from validated models. The CV-value is set to 0.1. This uncertainty only affects the Cs-concentration in fish.

6. The delay factor (BUD) influencing the biouptake of radiocesium in such a manner that fish larger than 100 g ww should be expected to have a slower biouptake than smaller fish, which are often planktivores and reach a dynamic steady-state within one month (the calculation time of the model). This is also an uncertain model variable. The CV is set to 0.5. This uncertainty only affects the Cs-concentration in fish.

7. The autochthonous factor (Y_{AU}), which accounts for "biological dilution" and is related to lake TP-concentrations. This factor is relatively uncertain, although the underlying effect of "biological dilution" is well documented and almost self-evident since the definition of the mean Cs-concentration in fish is the amount of radiocesium in fish divided by fish biomass. CV is set to 0.25. This uncertainty only affects the Cs-concentration in fish.

8. The allochthonous factor (Y_{AL}), which also accounts for "biological dilution" but not from increased phytoplankton production but from increased bacterial production from allochthonous carbon, and subsequent increases in biomasses of zooplankton and fish. This process is important in "brown-water" lakes. The CV is set to 0.25 also for this dimensionless moderator, which is calculated from the percentage of outflow areas (i.e., bogs and mires causing a transport of humic matter to the lakes). This uncertainty only affects the Cs-concentration in fish.

9. The potassium moderator (Y_K). This moderator is based on fundamental mechanistic processes regulating the competition between potassium and cesium uptake via cell-membranes. The CV is set to 0.25. This uncertainty only affects the Cs-concentration in fish.

10. The partition coefficient for cesium in lake water (Kd, or rather the constant 800,000 in eq. 1.26). The model for lake Kd is based on mechanistic processes in describing for how K-ions influence the binding of Cs-ions to carrier particles. There are, however, uncertainties concerning the constant 800,000 in this sub-model, and CV is set to 0.25.

11. The fraction of outflow areas (OA), which is used as a distribution coefficient. The value can be determined quite well from standard maps showing bogs, mires and upstream lakes. The CV is set to 0.1. This uncertainty can influence the secondary transport of radiocesium from land to water, and hence Cs-concentrations in water and fish.

12. The distribution coefficient regulating the transport from inflow areas directly to the lake and/or to outflow areas (D_{OAW}). The CV is set to 0.5.

13. The basic runoff rate from outflow areas (R_{Rd}). The initial value of 0.04 (i.e., a 4% loss during the first year after the fallout) emanates from several studies (see Håkanson et al., 1988; IAEA, 1999). These studies have also shown that this value is uncertain because many factors can influence the value, e.g., soil type, precipitation, season of the year and ice-conditions. CV is set to 0.25

14. The soil type or the soil permeability factor (SPF). This is an uncertain factor and there are major differences in soil types within and among catchment areas (soils always appear in a patchy manner, even within 1 m^2), so it is generally difficult to

assess what value should be used to characterise the entire catchment area. CV is set to 0.5.

15. The fall velocity (v_{Cs}). The value of the fall velocity is, together with the approach for determining Kd, the key regulator of the sedimentation process, and hence the amount of radiocesium in sediments, and also the internal loading. The v_{Cs}-value is derived from circumstantial information and not from direct measurements. CV is set to 0.5.

16. The concentration of suspended particulate matter (SPM) in the lake water, which influences the settling velocity of particulate cesium, the lake Kd-value, the deposition of matter onto the lake bed, and hence the age of the sediments, and the retention time of radiocesium in sediments. The values of SPM can be measured and/or predicted quite well and CV is set to 0.1.

17. The age of radiocesium on ET-areas (T_{ET}). If this age is set too short, say 1 week, the ET-areas would function as E-areas. If, on the other hand, the age is set too long, the ET-areas would function as A-areas. What value should be used? Generally, it is about 1 year for lakes and about 1 month for coastal areas, but this is an uncertain value and CV is set to 0.5.

18. The age of radiocesium on A-areas (T_A). This is determined by the sedimentation of matter, the definition of the thickness of the active layer and the bioturbation, all of which are uncertain factors. Thus, CV is set to 0.5.

19. The diffusion of radiocesium from A-areas (Diff). There are, as mentioned, some empirical data suggesting that a general default value for the diffusion rate can be set to 0.002. This value depends on the sedimentation of material, on the hypolimnetic temperature and bioturbation. CV is set to 0.5.

20. The ET-area (ET). This is a distribution coefficient in this model separating sedimentation of radiocesium on ET- and A-areas. The ET-areas can be measured and modelled quite well (see Håkanson and Jansson, 1983) and CV is set to 0.1.

21. The theoretical lake water retention time (Tw). This value is important. It is calculated from the inflow of water and it regulates the outflow of radiocesium from the lake (the openness of the "exit gate"). Tw is calculated from the area of the catchment, precipitation and surface runoff, which gives a value of the theoretical water discharge (Q), which together with lake volume gives Tw (Tw = Vol/Q). This is a standard lake variable in practically all contexts. CV is set to 0.1.

22. The exponent in the definition of the retention rate (Exp). The lake water retention rate is = $1/Tw^{Exp}$. Many tests have been done to obtain an adequate value and sub-model for this (see Håkanson and Peters, 1995). This is still a rather uncertain component of this model and CV is set to 0.25.

23. Last but not least, the fallout (Fall). It is evident that even though the fallout can be measured quite well, the value used for the fallout onto the lake and its catchment can be quite uncertain, especially for large catchments. This is the primary driving force for all subsequent calculations. The CV of the fallout has been set to 0.1.

The aim of the following tests is to produce a ranking of these uncertainties for the two target variables. It is evident that all uncertainties cannot be of equal importance for a given lake and that the ranking must differ among lakes

All the following tests use the data for Lake Hamstasjön (2217), one of the validation lakes listed in Table 1.4.

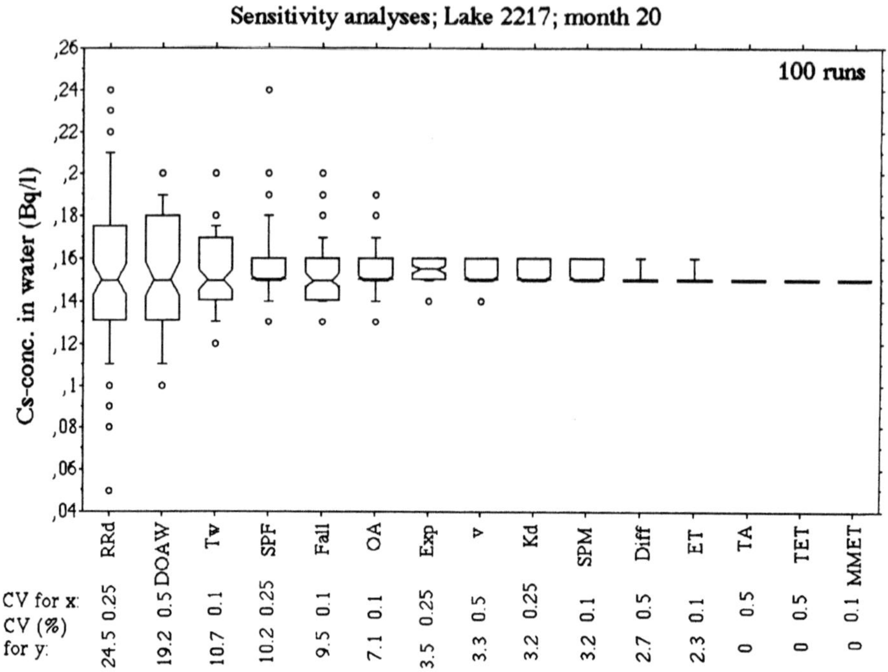

Fig. 1.38. Sensitivity tests for Cs-concentration in water for Lake Hamstasjön (2217), Sweden. These CV-values have been calculated for month 20 (month 1 is January 1986).

Figure 1.38 first gives the sensitivity tests for Cs-concentrations in lake water. The ranking is given by the CV-values for the Cs-concentrations in water and the box-and-whisker plots. The most important factor for the uncertainty in the Cs-concentrations in lake water is the runoff rate (R_{Rd}), which regulates the secondary flow of cesium from the catchment to the lakes. It should be noted that the factors influencing the secondary flow are very important in this lake: R_{Rd} is the most important factor, the distribution coefficient (D_{OAW}) is the next most important factor, the soil permeability factor (SPF) is fourth and the value for the outflow areas (OA) is fifth. Uncertainties in Tw, and hence the lake water retention rate, are also important, as well as uncertainties in fallout. The factors regulating sedimentation (the settling velocity, v_{Cs}, the lake Kd-value and the suspended particulate matter concentration, SPM), are also quite important, whereas the conditions in the sediments are less important. The ET-areas only occupy 15% of the area in this lake. In lakes more dominated by erosion and transport processes, one must expect other results from sensitivity tests.

Figure 1.39 gives the corresponding uncertainty tests. It should be noted that the total CV for Cs-concentrations in lake water is 37.2%. The CV for empirical data for within-lake variations in Cs-concentrations is 0.3 (see Table 1.2). The correspondence between the calculated CV of 0.37 and the empirical CV of 0.3 means that the CV-values selected in these uncertainty analyses for the model variables produce an appropriate calculated CV in the target variable. This is a critical control of the presuppositions for the test. Generally, one would apply the precautionary principle and use CV-values that are too high rather than too low for the model variables so that the

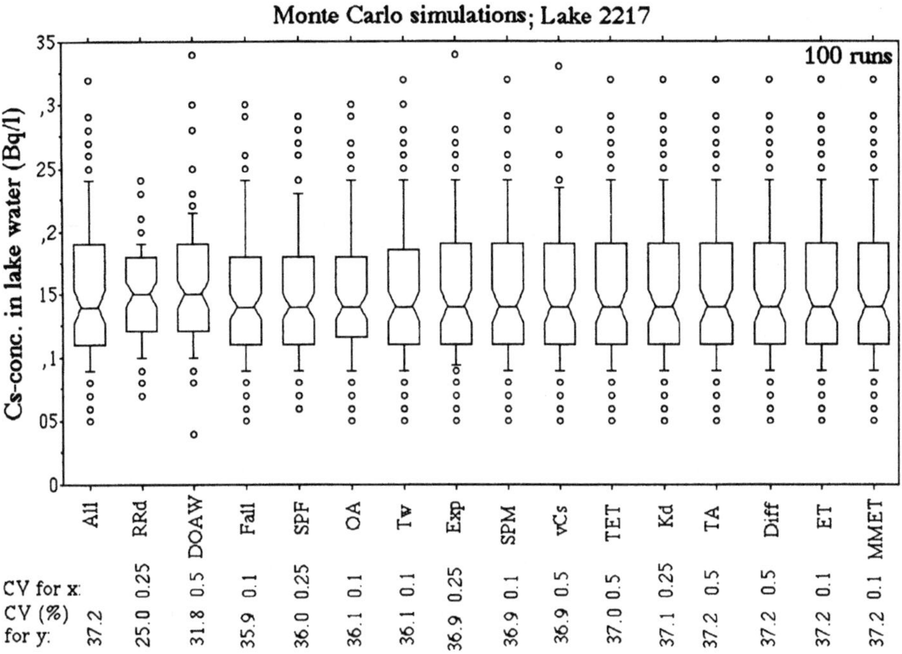

Fig. 1.39. Uncertainty tests for Cs-concentration in water for Lake Hamstasjön (2217), Sweden. These CV-values have been calculated for month 20 (month 1 is January 1986).

calculated CV for y is higher than the empirical value. This is also what these Monte Carlo simulations show.

In short, this result indicates that the assumptions concerning the CV-values for the model variables are reasonable.

The results of the Monte Carlo simulations agree quite well with the results from the sensitivity analyses. The most important factor is again the runoff rate (R_{Rd}). If this uncertainty is omitted, the CV for the Cs-concentration in water decreases the most, from 37.2% to 25.0%. The next most important factor also in this test is the D_{OAW}-value. This means that future model improvements should concentrate on getting more reliable data and/or sub-models for R_{Rd} and D_{OAW}. This would be the best way to reduce the uncertainties in the model predictions for this lake. It should be stressed again that, since the model is meant to be used generally, it must be well balanced for all lakes in the given model domain. Even though the values for the ages of ET- and A-sediments, the value used for the ET-areas and the diffusion rate were not so important in this lake, they will be of greater importance in shallower lakes. The uncertainty and sensitivity analyses are only indicative of how the model predicts. The ultimate test of predictive power is NOT sensitivity and uncertainty tests, but validation.

Next, results are given for Cs-concentrations in fish (here 1000 g pike). One would expect the calculated CV-values to be higher, not because the CV-value for empirical Cs-concentrations in fish is higher, which it is not (the characteristic CV for Cs in lake fish is 0.22 and for lake water 0.3; see Table 1.2), but because one expects it to be much more difficult to model and predict Cs-concentration in fish, since these

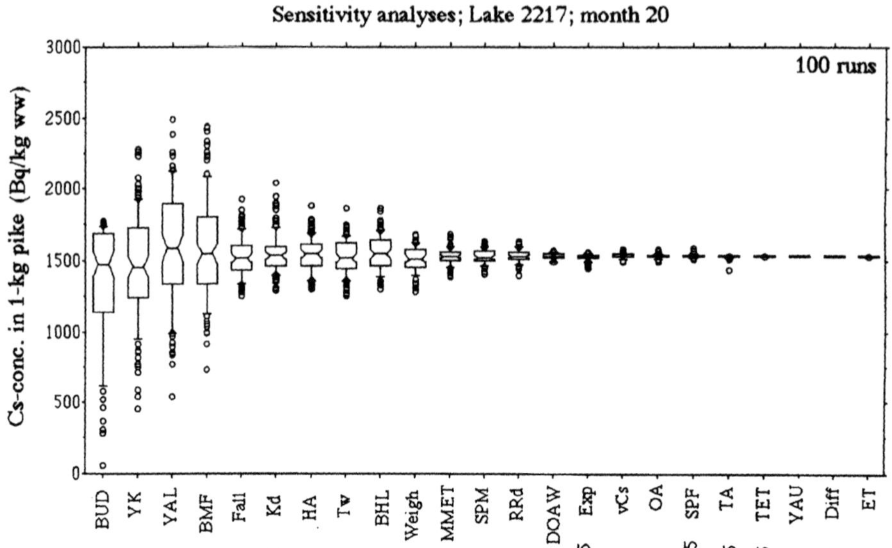

	BUD	YK	YAL	BMF	Fall	Kd	HA	Tw	BHL	Weigh	MMET	SPM	RRd	DOAW	Exp	vCs	OA	SPF	TA	TET	YAU	Diff	ET
CV for x:	0.5	0.25	0.25	0.25	0.1	0.25	0.5	0.1	0.1	0.2	0.1	0.1	0.25	0.5	0.25	0.5	0.1	0.25	0.5	0.5	0.25	0.5	0.1
CV (%) for y:	32.6	26.5	26.0	22.7	9.4	9.0	8.0	8.0	7.9	5.9	4.0	3.3	3.1	1.5	1.5	1.2	1.1	0.8	0.7	0.04	0	0	0

Fig. 1.40. Sensitivity tests for Cs-concentration in 1000 g pike for Lake Hamstasjön (2217), Sweden. These CV-values have been calculated for month 20 (month 1 is January 1986).

predictions incorporate predictions of Cs-concentrations in water as well as many other processes related to feeding habits, biouptake, etc.

The results are given in Figure 1.40 for sensitivity analyses, and Figure 1.41 for uncertainty analyses. These figures demonstrate the following points:

• The calculated CV for y has, as expected, increased from 37.2% for Cs-concentrations in lake water to 57.0%.

• The ranking of the factors contributing to this CV is quite similar in the two tests. The sensitivity test gave:

Delay factor (BUD) > K-moderator > Allochthonous moderator > BMF > Fallout > Sediment factor (HA) > Tw > BHL > Fish weight > Epilimnetic temperature.

The uncertainty tests gave:

K-moderator > Delay factor (BUD) > BMF > BHL > Kd > Tw > Allochthonous moderator > Sediment factor (HA) > SPM.

It is evident that the uncertainties associated with the biouptake delay factor (BUD), the potassium moderator and the feeding habits (BMF = biomagnification factor, BHL = biological halflife and sediment factor, HA) are key components in the fish sub-model.

• The model is well balanced. No part of the model totally dominates the calculated uncertainties in the two target variables.

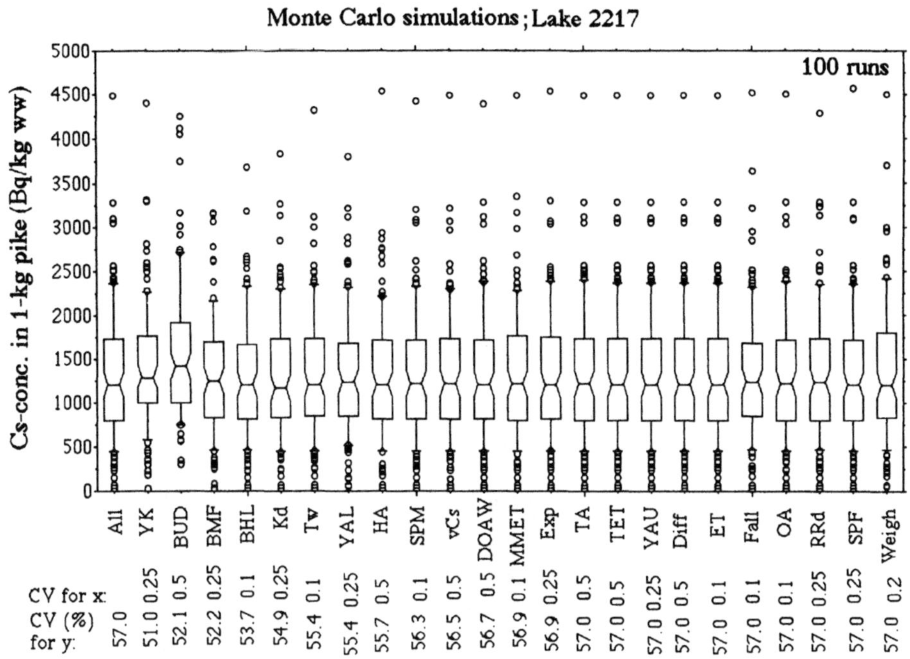

Fig. 1.41. Uncertainty tests for Cs-concentration in 1000 g pike for Lake Hamstasjön (2217), Sweden. These CV-values have been calculated for month 20 (month 1 is January 1986).

1.3.4.3. Validations

From the results given in figures 1.38-1.41, we will turn to the final model testing, i.e., validation. The validation lakes are given in Table 1.4. They cover a wide range of lake characteristics, although they are all European: sizes are from 0.041 km^2 to 1147 km^2, mean depths are from 2.9 to 90 m, characteristic pHs are from 5.1 to 8.5, trophic levels rage from oligotrophic (TP from 10 μg/l) to hypertrophic (TP = 60 μg/l), and fallout ranges from 2 to 130 kBq/m^2. The data base includes seven fish species.

The crucial question is: what can be expected in terms of predictive success?

1.3.4.3.1. Previous validations of comprehensive lake models

To ensure realistic expectations of model performances, it is worth explaining the background and focus on the situation about 10 years ago. At that time many models predicted concentrations of radionuclides and metals in fish with an uncertainty of a factor of around 10. The BIOMOVS-project (BIOsperic MOdel Validation Study; see, e.g., BIOMOVS, 1990) has been instrumental in disclosing poor ecosystem models, which is a key to improving model performance. The BIOMOVS-project showed that, e.g., model predictions of Hg in fish were about one order of magnitude above measured values, and that the uncertainty limits around this prediction and around those of the other models tested in that project were more than two orders of magnitude apart

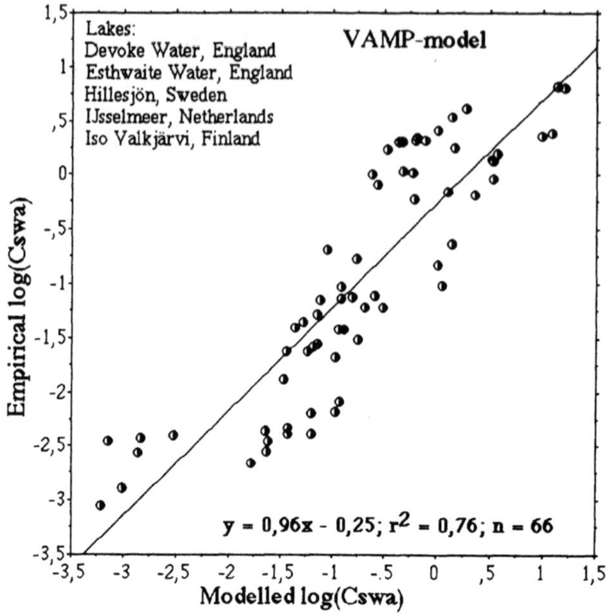

Fig. 1.42. Modelled values for Cs-concentrations in lake water versus empirical data (log-transformed values) for the VAMP-model.

(from 0.1 to 10!!). It is an understatement to say that such predictions are uncertain. Many ecosystem modellers today probably argue that good models yield uncertainties of a factor of around 2.

The VAMP-project (IAEA, 1999) has made a significant contribution to the evolution of predictive aquatic ecosystem modelling. This is shown in Figure 1.42, which illustrates how well the VAMP-model for lakes performed when tested for the five lakes for which data were available in that project (see Tables 1.3 and 1.4 for further information on these lakes). The VAMP-model did very well indeed, yielding an r^2-value of 0.76 and a slope close to the ideal (0.96). However, it must be noted that this good result is NOT obtained from validations but from calibrations. In any case, and this is important, the model variables were NOT altered but kept as model constants in obtaining the results in Figure 1.42. There was no "tuning" of the model, a well-known practice among modellers in the past. Thus, the VAMP-model can give good predictions. The VAMP-model is, like this model, a rather comprehensive dynamic model.

The interesting question now concerns the results for this model of validations not just for Cs-concentrations in water but also for fish. Cs-concentrations in fish are important in dosimetric models and because such concentrations represent biological aspects of radionuclide contamination of aquatic ecosystems, which Cs-concentrations in water do not. What can be expected in terms of predictive success?

1.3.4.3.2. Uncertainties in empirical data

One way of determining the highest possible r^2 for a predictive model is to compare two empirical samples (the Emp_1-Emp_2 test protocol, see Håkanson and Peters, 1995; IAEA,

1999). The variables in these two samples should be as time- and area compatible as possible; they should be sampled, transported, stored and analysed in the same manner. To illustrate the basic approach for determining the highest r^2 from Emp_1-Emp_2 comparisons, r_e^2, Table 1.9 gives a compilation of information from such comparisons for data of the VAMP lakes [for Cs-concentrations in water (in Bq/l), and in whitefish, trout, small perch, large perch, roach and pike (in Bq/kg ww)]. The first column gives r_e^2-values when the two empirical samples are compared (i.e., r_e^2, which should be as close to 1 as possible if the data are reliable), and the next column gives the slopes (which should also be close to 1) and then the number of data in the regressions (N). There are evidently major uncertainties in this data set, and also certainly in many other data sets. In fact, there are data-series yielding r_e^2-values of 0 (!), e.g., for large perch in Devoke Water and for Cs-concentration in lake water in Lake Hillesjön. The median r_e^2-value is 0.68, and there is no significant difference in the r_e^2-values for lake water and fish. This gives an interesting background to the results in Figure 1.42. It actually shows that the VAMP-model in some of these lakes predicts the mean monthly Cs-concentrations in lake water better than the measured data. This is interesting, and maybe a little provocative if one tends to regard empirical data as a solid base for calibrations and validations. There are empirical data which correctly disclose the conditions in the sample bottles or in the sample buckets, if several samples are pooled, but the data may provide a very poor representation of the mean monthly conditions in the lake, which is what the model predicts.

Yves Prairie (1996) has produced some very useful results illustrating the practical utility of models for predictions of individual y-values, here referred to as Prairie's "staircase". If the confidence bands are wide apart when modelled values are compared to empirical data, then the model can produce totally useless predictions for individual y-values. The usefulness of the predictions is directly related to the number of steps, or classes, obtained in the "staircase", and this is related to the given r^2-value, as illustrated in Figure 1.43. The number of classes (NC) is also determined by the statistics (the statistical certainty, p, and the number of data used in the regression, N). If the 95% confidence bands are used, then the relationship between the number of classes (NC) and the r^2-value obtained when empirical data are compared to modelled values is given by:

$$NC = 1.32/(1-r^2)^{0.5} \qquad (1.60)$$

Figure 1.43 gives the relationship between NC and r^2 for different N-values, and one should note that if $N > 6$, eq. 1.60 gives a good description of the relationship.

Figure 1.44 illustrates some other fundamental concepts related to the question of "the highest possible r^2". Figure 1.44A gives empirical data for the target variable y on both axes, i.e., Emp_1 vs Emp_2. There are uncertainties for all these values. One way of quantifying such uncertainties is by means of the CV-value. It is evident that the CV-value of a given variable e.g., chlorophyll-a concentration in the context of eutrophication or Cs-content in fish muscle in the context of radioecology, is NOT a

Table 1.9. Comparison in terms of r_e^2, slope and number of data in the regression (N) for two empirical data sets (Emp_1 and Emp_2) for all available Cs-concentrations in water and fish for the VAMP lakes. Modified from Håkanson (1997).

	Lake	**Country**	**Emp1 vs Emp2** r_e^2	**slope**	**N**
Water	IJsselmeer	Netherlands	0.884	3.020	4
Water	Iso Valkjärvi	Finland	0.951	0.706	9
Water	Devoke Water	England	0.113	-0.279	4
Water	Esthwaite	England	0.400	1.247	8
Water	Hillesjön	Sweden	0.016	0.146	14
Whitefish	Iso Valkjärvi	Finland	0.309	0.890	5
Trout	Heimdalsvatn	Norway	0.862	0.930	25
Trout	Devoke Water	England	0.274	0.550	22
Smelt	IJsselmeer	Netherlands	0.915	0.706	9
Small perch	Hillesjön	Sweden	0.949	1.171	5
Small perch	Iso Valkjärvi	Finland	0.716	1.351	12
Small perch	IJsselmeer	Netherlands	0.852	1.804	9
Ssmall perch	Devoke Water	England	0.017	-0.195	6
Roach	Hillesjön	Sweden	0.653	1.034	7
Pike	Hillesjön	Sweden	0.854	0.738	4
Pike	Iso Valkjärvi	Finland	0.440	0.473	5
Large perch	Hillesjön	Sweden	0.867	0.772	3
Large perch	Devoke Water	England	0.000	0.009	9
	Min.		0.00	-0.28	3
	Max		0.95	3.02	25
	MV		0.56	0.84	9
	M50		0.68	0.76	8

constant but a variable. The CV-value for within-ecosystem variability is always related to very complex climatological, biological, chemical and physical conditions, which means that simplifications are often requested. "Everything should be as simple as possible, but not simpler", according to Albert Einstein. We have tried to follow that advice here in building this model for radiocesium. It has been shown (see Håkanson and Peters, 1995) that it is often possible to define a characteristic CV-value for a given variable (see Table 1.2), e.g., 0.3 for Cs-concentration in water and 0.22 for Cs-

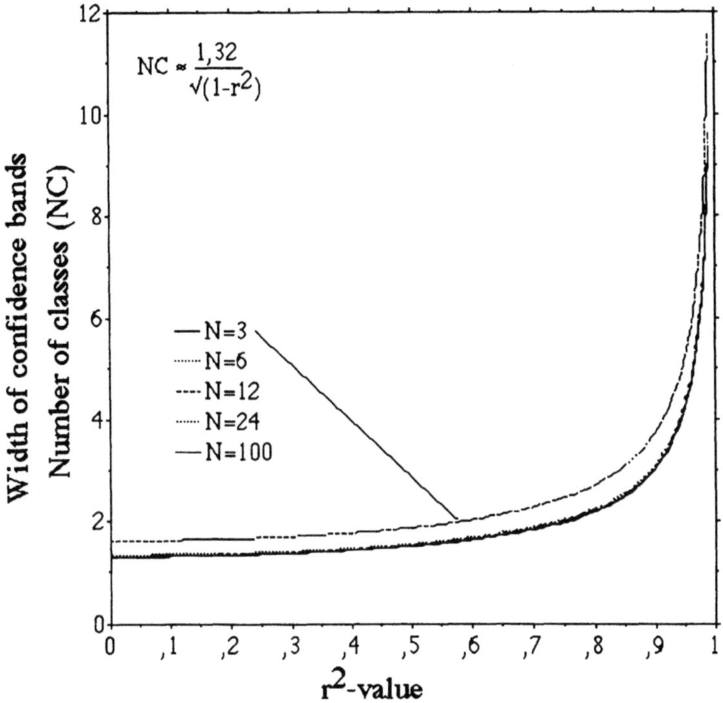

Fig. 1.43. The relationship between the number of classes (NC) and the r^2-value when modelled values are compared to empirical data, as given by Prairie's "staircase".

concentration in fish. Such characteristic CV-values are, evidently, very useful, e.g., in uncertainty analysis of models using Monte Carlo simulations when one wants to know typical ranges of model uncertainty.

Figure 1.44 illustrates a case in which a CV of 0.35 has been used for all empirical data on both axes since all data apply to one target y-variable. The uncertainty associated with the given target variable is illustrated by the uncertainty bands. This uncertainty will evidently influence the result of the regression, such as the r^2-value and the confidence intervals. If CV for y is large, one CANNOT expect a model to predict y well. Figure 1.44B illustrates a normal model validation when modelled values are put on the x-axis. The empirical uncertainty associated with y remains the same on the y-axis but the uncertainty in the x-direction is related to the uncertainty associated with the model structure and the uncertainty of the model variables (x). Generally, one would expect the model uncertainty to be larger, or much larger, than the uncertainty in the y-direction. This is also illustrated in Figure 1.44B.

This means that the r^2-value in the regression in Figure 1.44B is likely to be lower than the r^2-value obtained in the Emp_1-Emp_2 comparison. The residual value, $R^2 = 1-r^2$, must then be higher. The crucial question is: how far is it possible to reduce this residual uncertainty? This can only be achieved by building/structuring the model in the best possible manner by allowing for the most important processes/model variables

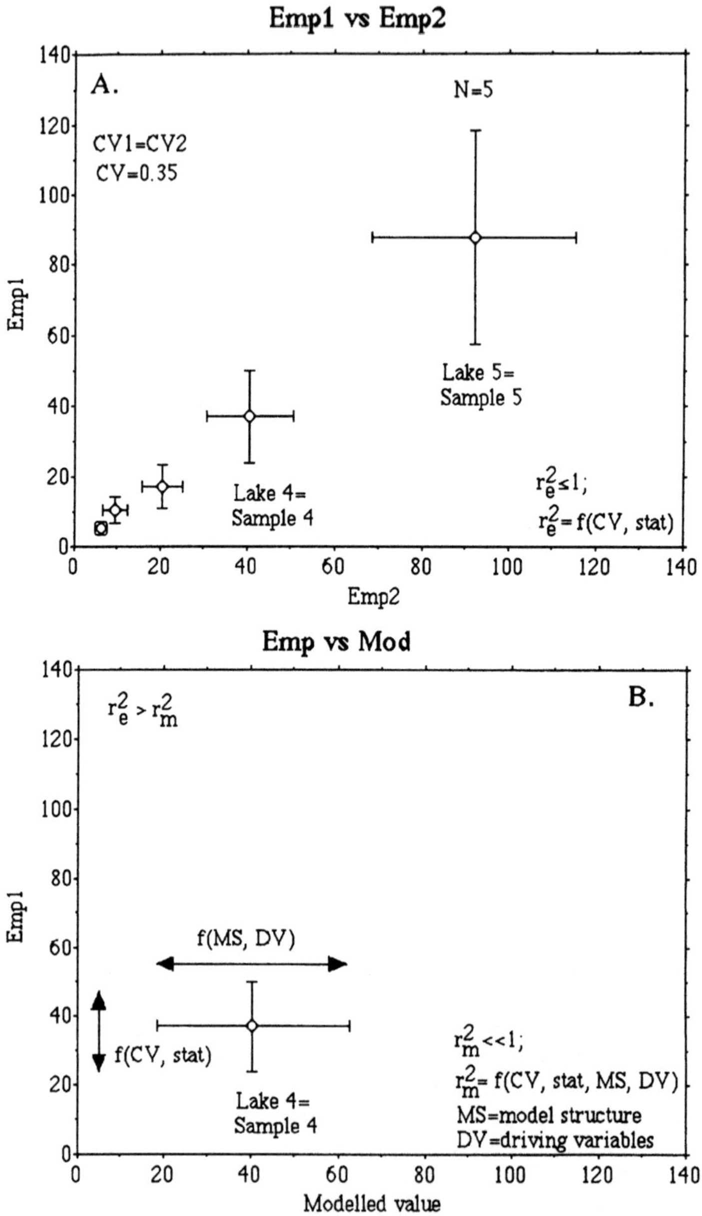

Fig. 1.44. Illustration of some fundamental concepts related to the question of "the highest possible r^2" of ecosystem models.
A. Empirical data for the target variable y on both axes, i.e., Emp_1 vs Emp_2.
B. Empirical data on the y-axis and modelled values on the x-axis.

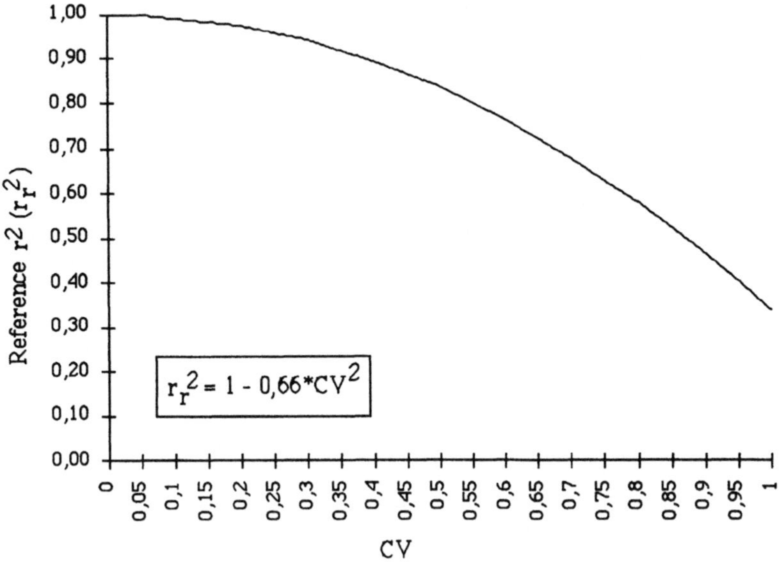

Fig. 1.45. The relationship between "the highest reference r^2" (r_r^2) and the characteristic coefficient of variation for variability within (CV) ecosystems.

and by omitting the relatively unimportant processes/variables which can add more to the model uncertainty (i.e., add to the width of the uncertainty bands, CI), than to the predictive success (i.e., increase the r^2-value; see Håkanson and Peters, 1995; Håkanson, 1995 for a more detailed discussion on optimal model size).

Figure 1.44 illustrates a rather simple scenario for empirical/statistical regression models that produce one y-value for one ecosystem (e.g., a lake). In this context, one must also discuss these matters from a general perspective and then include dynamic models, which yield time-dependent predictions of y, i.e., time series of y, where the data in the time series are not independent of each other.

Håkanson (1999) has derived a simple, general expression for the highest reference r^2, r_r^2, related only to the characteristic CV for the target y-variable in a predictive model:

$$r_r^2 = 1 - 0.66 \cdot CV^2 \tag{1.61}$$

The equation is graphically shown in Figure 1.45.

The important message in Figure 1.44 is that the number of classes increases very rapidly for r^2-values higher than about 0.75, and models yielding r^2-values lower than that are more or less useless for predictions of individual y-values (but not necessarily for mean y-values in regional modelling for which incorrect predictions for individual lakes can be accepted; see Håkanson, 1991). The message from Figure 1.45 is than one can never expect models to predict better than the highest reference r_e^2, which is related to the empirical uncertainty in the y-variable to be predicted.

From this, one can conclude that the VAMP-model represents a great step forward in the modelling of contaminants, e.g., compared to many of the models used in the BIOMOVS-project, and that the VAMP-model predicts better than "poor" empirical data but "worse" than "good" empirical data.

How well, then, can this model predict?

1.3.4.3.3. The validation of the lake model for radiocesium

The validations of the radiocesium model presented in this section have been done in the following manner.

1. The validation lakes were selected to cover a wide domain (see Table 1.4).

2. Empirical data were first directly compared to modelled values according to the procedure illustrated in Figure 1.46. Figure 1.46A gives the empirical data for Cs-concentration in 500 g pike in Lake Siggefora and the corresponding modelled values. These data were then regressed against one another, as illustrated in Figure 1.46B. The r^2-value in this case was 0.51, and the number of fish was 22. This regression only covers one lake and one type of fish and, hence, has a limited range. How does the model predict for all the given species of fish in all the validation lakes? Also note that the modelled data are not independent of one another and that the frequency distribution of the data on the y- and x-axis are not normal.

3. All data-series like the one given in Figure 1.46 for 500 g pike in Lake Siggefora have been compiled for lake water and for all species of fish. This is also the way in which data are presented in Figure 1.42.

The results for the Cs-concentrations in lake water are given in Figure 1.47. The r^2-value is 0.923(!). The figure gives the 95% confidence intervals for the mean y and the individual y and the regression line. The slope is 1.087. This regression is based on 61 data covering a range from 0.001 to 4.5 Bq/l. This is, of course, an amazing result for a validation. It actually means that the model predictions are as good as carefully conducted measurements - under the given conditions and in the given domain of the model.

Figure 1.48 gives the corresponding results for Cs-concentrations in lake fish. The data-set comprises 185 values for seven species of fish covering a very wide range, from 2 to over 30,000 Bq/kg ww. There are three very typical outliers. If these three outliers are omitted, the r^2-value is 0.98. This is almost like an analytical solution. If the three outliers are included, the r^2-value is 0.95. The slope is almost perfect. The r^2-value for the actual data is 0.86. This demonstrates, even proves, that the structuring of this model is very good. The model gives better predicted values than would often be obtained from sampling in standard lake-monitoring programs. If almost "perfect" predictions can be obtained for radiocesium for fish in lakes, this holds great promise for other types of contaminants in future models. It also means that models of this type in the future are likely to become widely used and excellent tools in remedial contexts.

It should also be stressed that there are several reasons why one should use log-transformed values in regressions of this type. One such argument concerns the frequency distributions. The following data show that Cs-concentrations in lake water, lake fish and suspended particles are log-normally distributed:

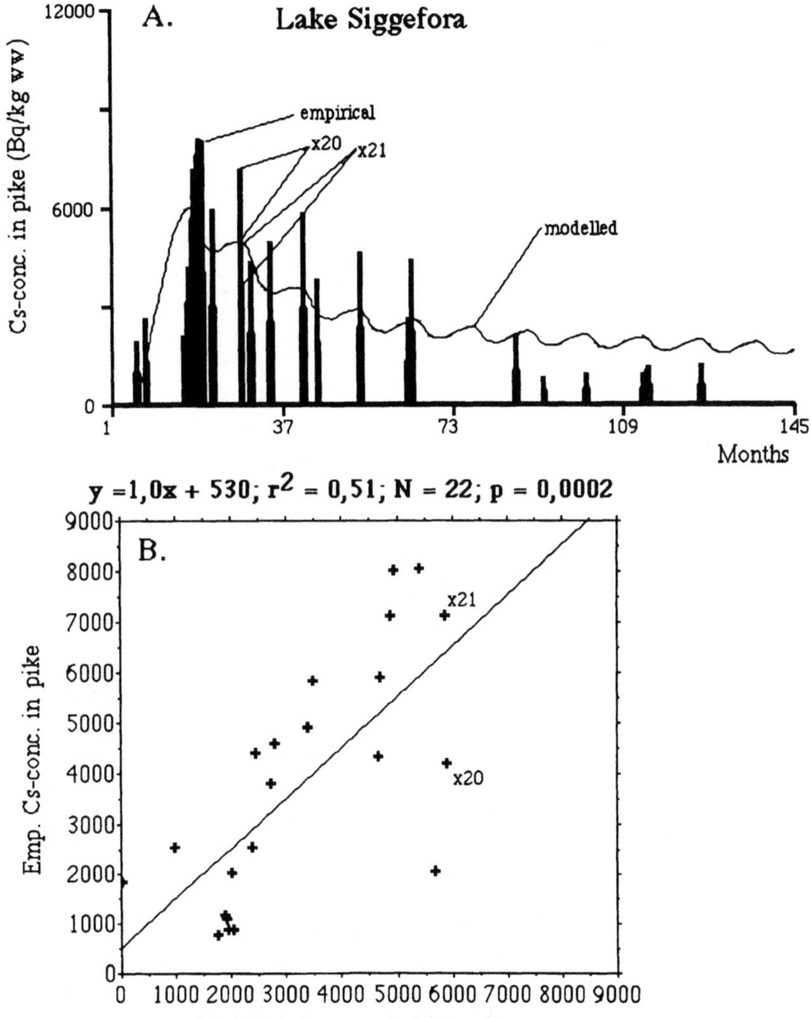

Fig. 1.46. Illustration of the relationship between empirical data and modelled values for 500 g pike in Lake Siggefora. (A) gives a direct comparison on a time-scale; (B) gives the corresponding regression.

	MV/M50-ratio		
	Actual	Log	N
Empirical, fish	1.80	1.87	186
Modelled, fish	2.18	0.89	
Empirical, water	7.90	0.95	63
Modelled, water	3.00	1.22	
Empirical, particulate	4.37	1.08	19
Modelled, particulate	7.01	1.16	

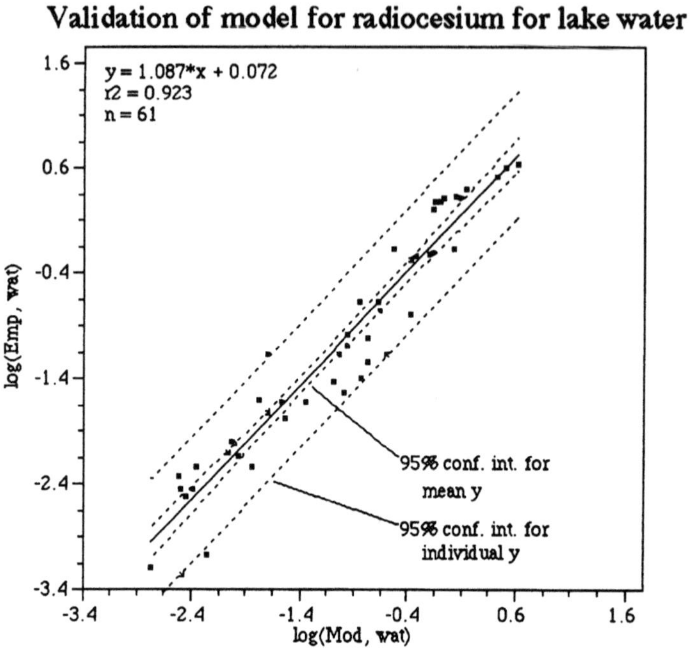

Fig. 1.47. Validation of the radiocesium model for lake water.

Some of these distributions are very positively skewed with MV/M50-ratios higher than 3. The log-transformation will produce more normal frequency distributions and MV/M50-ratios closer to 1. This is a requirement in regressions.

The list of MV/M50-ratios also includes Cs-concentrations on suspended particulate matter. The fact that such data are available opens up an interesting testing opportunity, i.e., a blind test. The model has not been calibrated to predict Cs-concentrations on suspended particulate matter, but it is easy to calculate those concentrations as $1000 \cdot (C_{tot}-C_{diss})/SPM$ (C_{tot} is total Cs-concentration in Bq/l; C_{diss}, is dissolved Cs-concentration in Bq/l and SPM is suspended particulate matter concentration in mg/l). This means that one can test how well the model predicts these Cs-concentrations. There are data available from one of the validation lakes, Lake Siggefora. The results are given in Figure 1.49. The r^2-value for the actual data is 0.94; it is 0.80 for the logarithmic data.

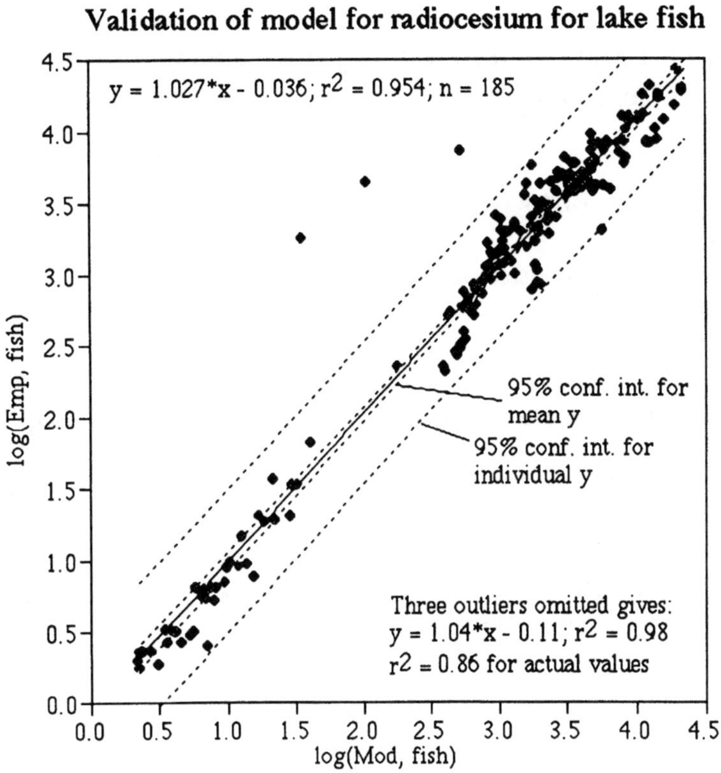

Fig. 1.48. Validation of the radiocesium model for lake fish.

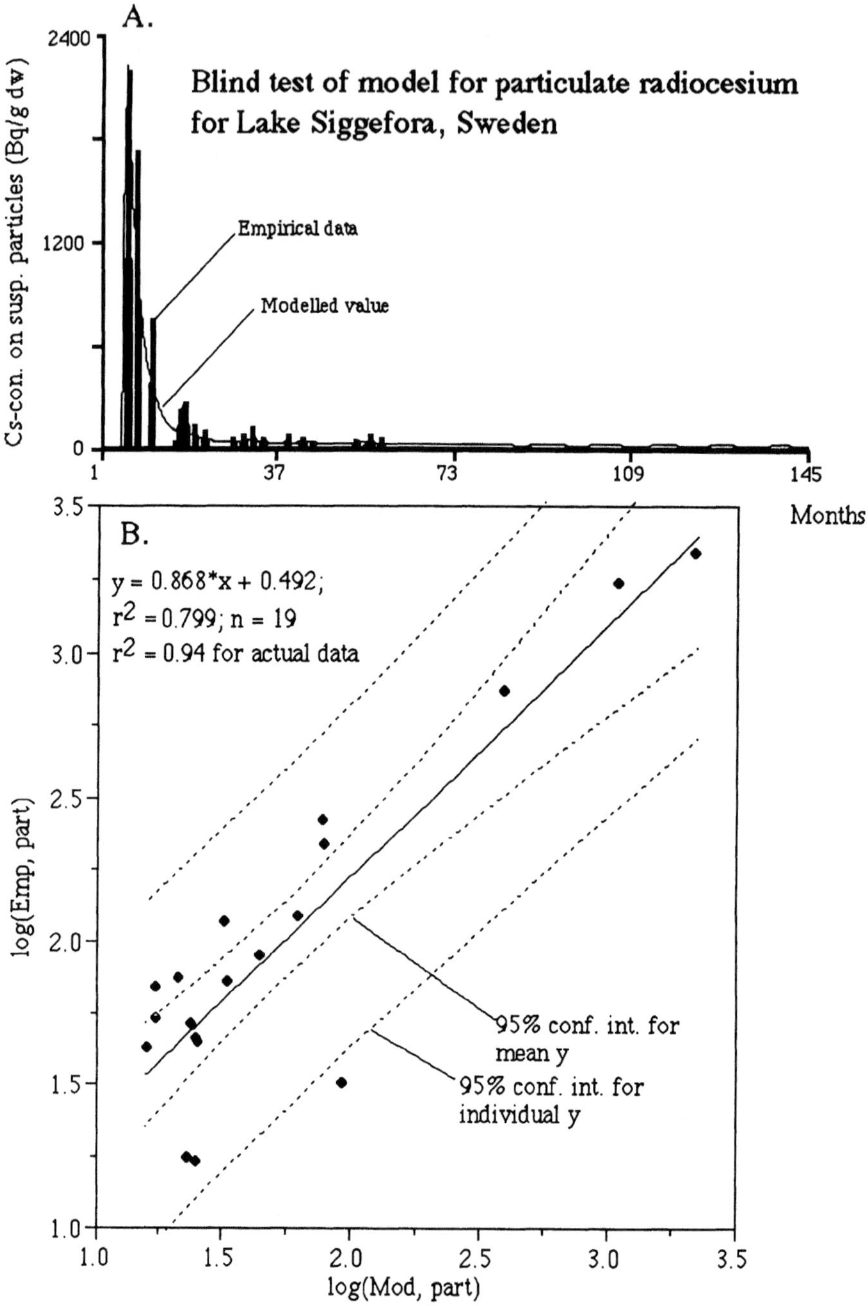

Fig. 1.49. A blind test of the radiocesium model for suspended particulate matter. (A) gives a direct comparison between modelled values and empirical data; (B) gives the regression.

CHAPTER 2

MODELLING RADIOCESIUM IN COASTAL AREAS

2.1. Aims

The following model for radiocesium has been constructed as a tool for quantifying the most important processes regulating the spread and biouptake of metals and radioisotopes in coastal ecosystems more generally.

2.2. Coastal Processes

2.2.1. DEFINITIONS

It should be stressed that it is NOT possible to make simple adjustments to lake models for marine areas because lakes and coastal areas differ in several respects.

• The hydrodynamical conditions are much more dynamic in coastal areas (a typical surface water retention time is 2-4 days for a coastal area and about 1 year for a lake; see Håkanson, 1999). This has implications for the conditions in the coastal areas, which are evidently greatly influenced by the conditions in the outside sea and/or adjacent coastal areas in direct contact with the sea.

• The amount of suspended particulate matter (SPM) always depends on two main causes, allochthonous inflow and autochthonous production. In the Baltic, however, there is also another primary source, land uplift (see Figure 2.1). Thousand-year-old sediments influence the Baltic ecosystem today. When the old bottom areas rise after being depressed by the glacial ice, they will eventually reach the wave base, which is the water depth above which the waves can exert a direct influence on, and resuspend, the sediments. It is easy to imagine that storms leading to high waves have a major influence on coastal ecosystems. The wave base, or the critical depth separating bottom areas where transport processes dominate the bottom dynamic conditions from areas of continuous sedimentation of fine materials, depends on the effective fetch (Figure 2.2), and the duration and velocity of the winds; during storms, the wave base may reach water depths of about 50 metres in the Baltic Proper (outside the coastal zone; see Håkanson, 1999). So, as a result of land elevation, the old sediments deposited hundreds and thousands of years ago will be resuspended. The land uplift in the Baltic varies from about 9 mm/yr in the northern part of the Bothnian Bay to about 0 for the southern part of the Baltic (Figure 2.1). This implies that large amounts of old glacial and post-glacial sediments are eroded. In this way, the carbon, nitrogen and phosphorus contained by these sediments, as well as metals and mineral particles, will again enter the ecosystem of the Baltic, perhaps thousands of years after they were originally deposited onto the bottom in a considerably calmer environment. Recent measurements demonstrate that as much as 80% of the material sedimenting onto the deep bottoms may be old eroded material (see Jonsson, 1992). The new supply of material to the sea from rivers, direct discharges and the bioproduction in the system, only contribute about 20% of the amount annually deposited on the accumulation areas (A-areas). The

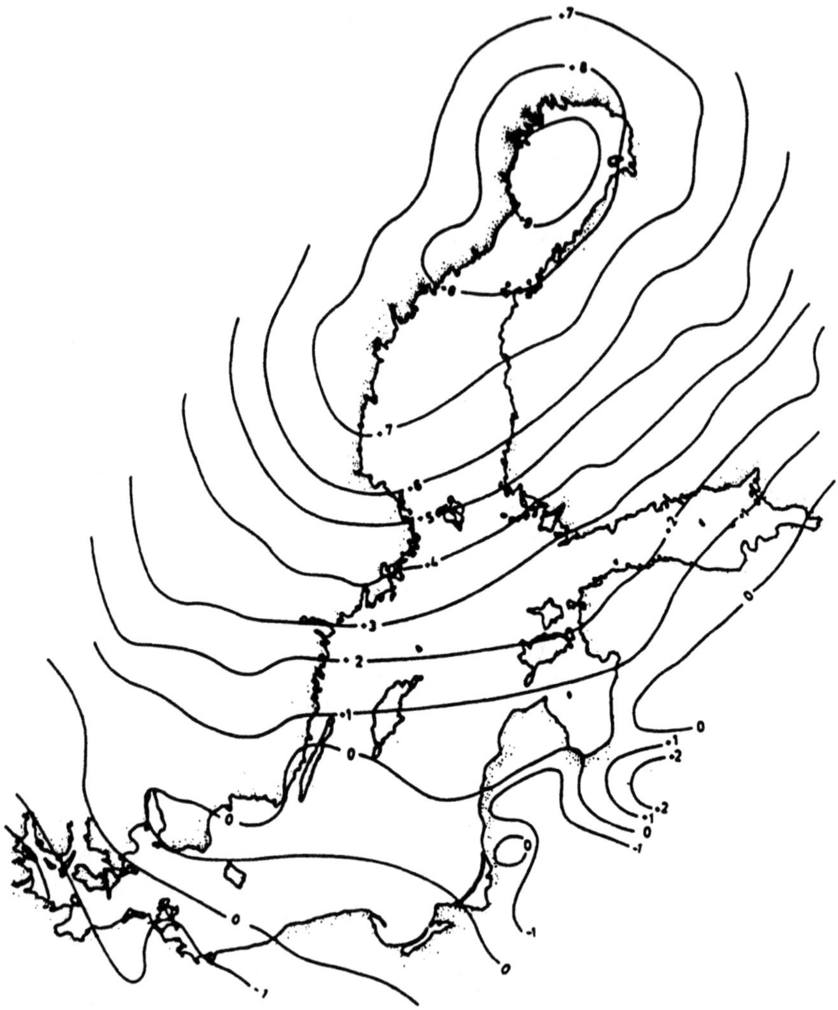

Fig. 2.1. Present-day land uplift in the Baltic Sea region. Values in mm/yr. Modified from Voipio (1981).

amount of matter deposited on the areas of erosion and transport (ET-areas) may be resuspended by, e.g., wind/wave action or slope processes, so resuspension is the fourth main process influencing the flux of matter in coastal areas. Resuspension is, however, a secondary process since it depends on the three primary processes. Land uplift is not a major primary source of matter in most lakes and coastal areas, but is, as has been stressed, a very important process in the Baltic.

• The allochthonous inflow of matter to a lake is simply $Q \cdot C_{in}$ (Q = tributary water discharge in m^3/month); C_{in} = tributary concentration of the given pollutant in g/m^3 or Bq/m^3). For coastal areas, the tributary inflow of matter is, by definition,

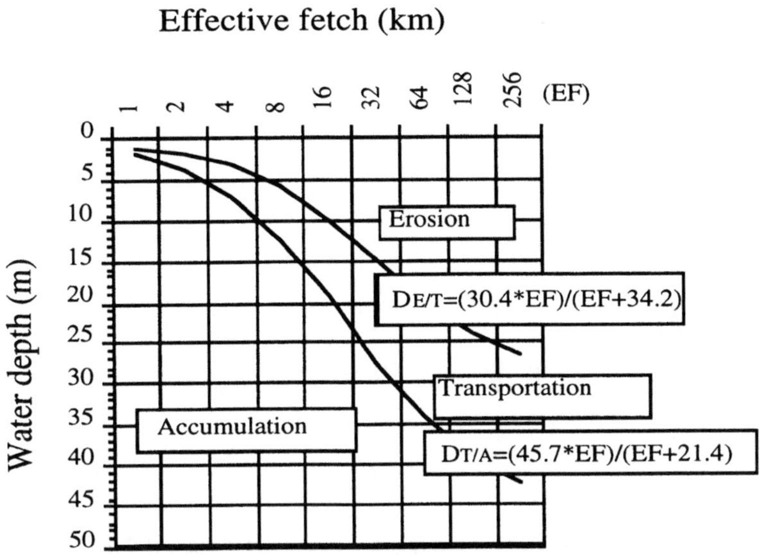

Fig. 2.2. The ETA-diagram giving the relationship between the effective fetch (the free water surface over which winds influence waves), the water depth and the potential bottom dynamic conditions. $D_{E/T}$ is the water depth separating E- and T-areas. $D_{E/T}$ can be predicted from the given equation. Modified from Håkanson (1999).

important in estuaries, but outside the estuaries the inflow is given by $R \cdot C_{sea}$, where R is the net inflow of water from the sea and/or adjacent coastal areas (m^3/month) and C_{sea} is the concentration of the pollutant in the water outside the coast (in g/m^3 or Bq/m^3).

• The autochthonous production in a lake can be predicted quite well from information on mean values for total phosphorus (Schindler, 1977, 1978; Chapra, 1980; Boers et al., 1993). For coastal areas one must account for the nitrogen (Redfield, 1958; Ryther and Dunstan, 1971; Ambio, 1990; Nixon, 1990).

• The bottom dynamic conditions in coastal areas also differ from those in lakes. In defining the bottom dynamic conditions (erosion, transportation and accumulation), this model uses the definitions from section 1.

The bottom dynamic conditions influence, and are influenced by, the hydrodynamic conditions, mixing, stratification, wind influences, etc. In sheltered coasts, one can find A-areas even at relatively small water depths, but coastal areas generally have smaller percentages of A-areas and larger percentages of ET-areas than lakes. This has profound effects for the internal loading and many internal transport processes of water pollutants.

• The salinity of coastal areas influences some major processes, such as stratification, and hence also mixing and retention. The salinity is of paramount importance to the number of species, as illustrated in Figure 2.3. It influences the aggregation of suspended particles, which is of particular interest in this model. The saltier the water, the greater the flocculation. This will be mathematically handled in this model by means of a dimensionless moderator for salinity, which influences the settling velocity of suspended particulate matter. This has a bearing on the sedimentation in coastal areas.

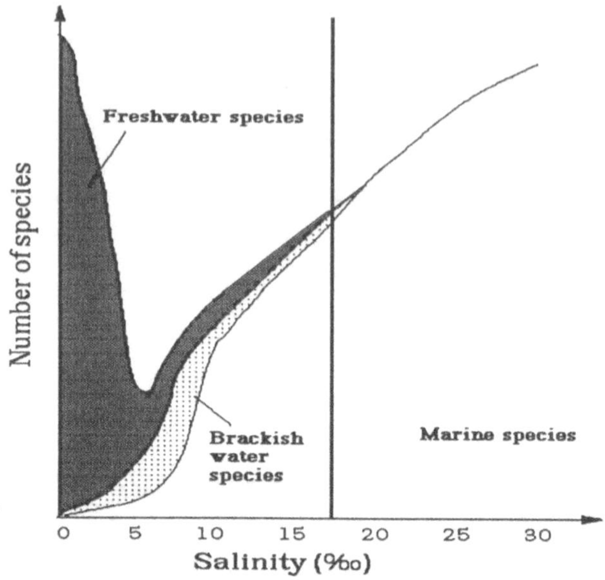

Fig. 2.3. The relationship between salinity and number of species. Redrawn from Remane (1934).

• A typical deposition in lakes is about 0.4 cm/yr (Håkanson and Peters, 1995). Of course, the value varies among and within lakes. Highly eutrophic lakes may have a net deposition of more than 1 cm/yr. Within lakes, the deposition generally increases from the wave base to the maximum depth (= sediment focusing; see Håkanson and Jansson, 1983). The deposition can be very large in limited deep-holes, say several cm/yr. In coastal areas, one generally expects much higher deposition than in lakes, especially in Baltic coastal areas. There are several reasons for this, one being related to the coastal currents (see Figure 2.4).

The dominant water circulation in each basin (the Bothnian Bay, the Bothnian Sea and the Baltic Proper in Figure 2.4) constitutes an anti-clockwise cell, which distributes the settling particles, the suspended material and the pollutants in a typical pattern, reflecting the flow of the water. This anti-clockwise cell is created by the rotation of the earth (the Coriolis force), which deflects any plume of flowing water to the right in relation to the direction of the flow in the northern hemisphere (and to the left in the southern hemisphere). Thus, when a Swedish river enters the Baltic, the water turns to the right and follows the shore.

1. The net hydrological flow is to the south on the west (Swedish) side of the Baltic.
2. The currents are rather strong and stable close to land and weaker towards the centre of each basin.
3. The figure illustrates only the net component of the flow; this means that the water would also flow in most other directions during the year.

The existence of these Coriolis-driven currents implies that relatively little material is transported to the deep open parts of the main basins since these currents

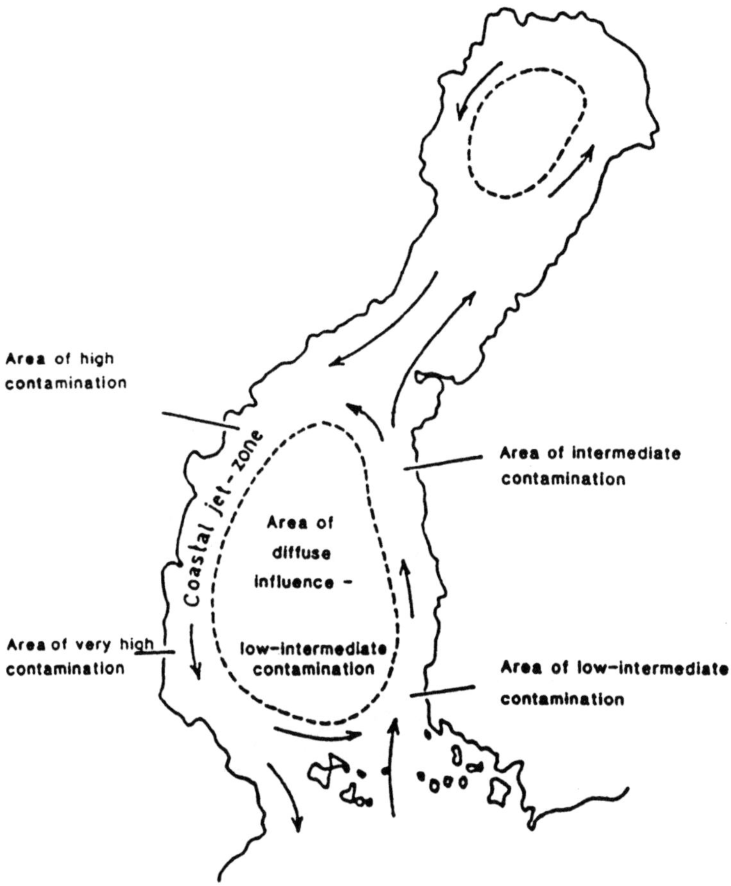

Fig. 2.4. Illustration of the coastal jet-zone and the major hydrological flow pattern in the Bothnian Bay and the Bothnian Sea. Modified from Håkanson et al. (1988b).

move the suspended particulate materials along and into the coasts. This also means that allochthonous matter from the large rivers entering the sea, the autochthonous production in the coastal areas and the resuspension in the shallow coastal areas together with the dominating coastal currents create an environment of high sedimentation within coastal A-areas. There are reported cases of deposition of more than 10 cm/yr (!) from the Stockholm archipelago (Markus Meili and Per Jonsson, pers. comm.), and this agrees well with the maximum value obtained using the model presented by Håkanson (1999) of 9.2 cm/yr. The average deposition value for Baltic coastal areas is 2-3 cm/yr, which is a factor of 5-10 greater than the characteristic value for lakes (0.4 cm/yr).

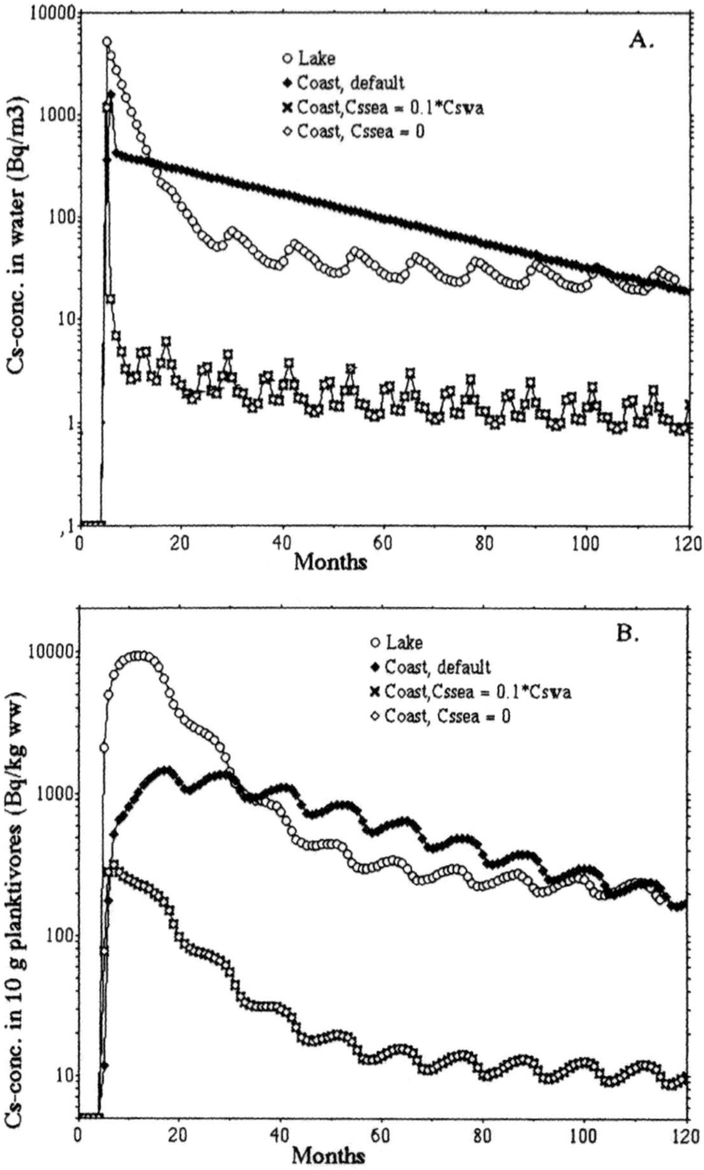

Fig. 2.5. Lake versus coast. A comparison (note the logarithmic scale of the y-axis) between the predicted Cs-concentrations in a lake and a coastal area in (A) water and (B) 10 g ww planktivores. The theoretical water retention time is 1 year for the lake and 0.16 months for the coastal area, while all else is equal.

It is evident that coastal systems, just like lakes, are extremely complex. It is also a matter of fact that there are very few validated models for chemical pollutants for coastal areas. One reason for this is, probably, that most scientists working in coastal and marine environments are not used to applying an ecosystem perspective, an approach that comes naturally to limnologists.

Figure 2.5 gives a comparison (note the logarithmic scale of the y-axes) between the predicted Cs-concentrations in a lake (using the model presented in section 1) and the model in this section for (A) water and (B) 10 g ww planktivores, if all the many characteristics for the lake and the coastal area are comparable, e.g., area (5 km^2), mean depth (9 m), size of catchment, catchment characteristics and bottom dynamic conditions (A-areas and ET-areas). The main difference is the theoretical water retention time, which is 1 year for the lake and set to 0.16 months for the surface water for the given coastal area.

Also note that concentrations of pollutants in coastal areas depend very much on the conditions outside the given coastal area, which are, evidently, very difficult to predict using a general, practically useful management model since there are great differences among coastal areas in this respect, because of oceanographic conditions, topographical presuppositions, latitude, tidal characteristics, etc. This coastal modelling approach uses three standard methods to account for the transport of radiocesium from the sea and/or adjacent coastal areas:

(1) a default approach based on fallout and an ecological halflife for radiocesium in the sea (which will be presented later);

(2) a simplified approach whereby the concentration of radiocesium in the sea (Cs_{sea}) is set at 10% of the calculated concentration in the coast, Cs_{wa}, because of a higher dilution in the sea,; and

(3) when Cs_{sea} is set to 0, which is an ultimate limit that would be unrealistic for many coastal areas, since it would mean, e.g., that there was fallout to the coastal area and the catchment but not to the sea outside the coastal area.

These different presuppositions for Cs_{sea} are, as indicated by the results in Figure 2.5, very important for the predicted Cs-values at the coast, which are also compared to analogous values for the reference lake. There are, however, no major differences if $Cs_{sea} = 0$ or $Cs_{sea} = 0.1 \cdot Cs_{wa}$. One should note that there are much higher predicted values if the default assumption given in this section is used. This means that the default conditions intentionally yield rather high values. The idea with this is to use the "precautionary principle" and not underestimate the consequences of a given fallout.

2.2.2. BASIC HYDRODYNAMIC PRINCIPLES AND PROCESSES FOR COASTAL AREAS

A coastal area may de defined and characterised in many ways, e.g., according to territorial boundaries, pollution status, water stratification (thermoclines/haloclines, see later for explanation), etc. One fundamental and broad way of characterising the entire system is according to geographical zonation into the following areas (see Figure 2.6).

1. The drainage area, also called the catchment area or, in American literature, the watershed. The rain falling on this area will, in due course, find its way to the open water areas. The drainage area of the Baltic covers 1,700,000 km^2, which is more than four times larger than the entire water area (415,266 km^2). Because of the climatological and geographical differences between the catchment areas of the different rivers, the water transport (and the chemical characteristics of the water) is very different in different rivers. There are significant seasonal variations in the river discharge. The maximum runoff generally occurs in the spring, during the thawing period.

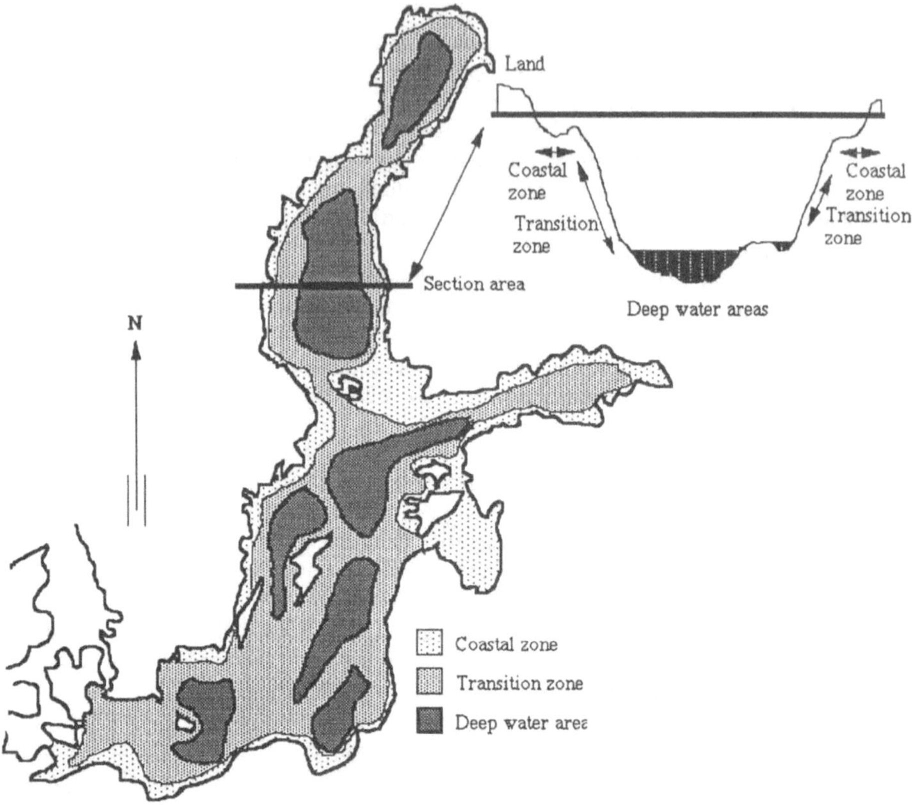

Fig. 2.6. The Baltic Sea may be divided into the following three functional zones: the coastal zone, the transition zone and the deep water area. Modified from Håkanson (1990).

2. The coastal zone, the zone inside the outer islands of the archipelago and/or inside barrier islands. This is the zone in focus in this model for radiocesium. The retention time of the water and the characteristics of the different types of pollutants may vary significantly between coastal areas. The coastal zone is of special importance for recreation, fishing, water planning and shipping and is a zone where different conflicts and demands overlap. The natural processes (water transport, flux of material and energy and bioproduction) in this zone are of utmost importance for the entire sea. It may be considered a "pantry and a nursery" (Håkanson and Rosenberg, 1985).

3. The transition zone, the zone between the coastal zone and the deep water areas. This is by definition the zone extending down to depths at which episodes of resuspension of fine material occur in connection with storm events and/or current activities (at about 50 m water depth in the Baltic; see Figure 2.7). The conditions in terms of water dynamics, distribution of pollutants (like nutrients, metals, radionuclides and chlorinated organics), and suspended and dissolved materials in this zone are of great importance for the ecological status of the entire system. This zone geographically dominates the open water areas outside the coastal zone in the Baltic.

4. The deep water zone, by definition the areas beneath the wave base. In these areas, there is a continuous deposition of fine materials. It is the "end station" for many types of pollutants and these are the areas in which conditions with low oxygen concentrations are most likely to occur.

Figure 2.7 illustrates why it is important to know the bottom dynamic conditions in contexts dealing with the spread of pollutants from a point source (in this example from the paper and pulp mill at Norrsundet on the Swedish coast in the Bothnian Sea) or from a river mouth. The fine materials (clays, humus, seston, etc.), which have a strong capacity to bind various types of pollutants, like radionuclides, are transported via several resuspension cycles from the site of emission to the true accumulation area at water depths greater than the wave base at about 50 m in the Bothnian Sea. Other coastal areas have, in contrast to the area outside Norrsundet in Figure 2.7, well-defined A-areas, and then the concentrations of pollutants emitted from a given point source increase markedly towards the site of emission (see Håkanson, 1999).

Many factors influence the water exchange in coastal areas (Figure 2.8). Emissions of nutrients or toxins cannot be calculated into concentrations without knowledge of the water retention time. If concentrations cannot be predicted, it is also practically impossible to predict the related ecological effects. Thus, it is important to introduce some basic concepts concerning the turnover of water in coastal areas.

The water exchange varies in time and space in any given coastal area. It can be driven by many processes, which also vary in time and space. The importance of the various processes will vary with the topographical characteristics of the coast, which do not vary in time, but vary widely between different coasts. The water exchange sets the framework for the entire biotic spectrum; the prerequisites for life are quite different in coastal waters where the characteristic retention time varies from hours to weeks.

Factors influencing the water exchange are listed below.

• The fresh water discharge (Q is often given in m^3/sec) is the amount of water entering the coast from tributaries per time unit. In small bays with large tributaries (estuaries), the Q-factor may be the most important factor for the water retention time.

• Tides. When the tidal variation is larger than about 50 cm, it is often a dominating factor for the surface water retention time. The tidal range is only about 2 cm in the Baltic.

• Water level fluctuations always cause a flux of water. These variations may be measured with simple gauges. They vary with the season of the year and are important for the water retention time of shallow coastal areas. Thus, the mean depth is a useful coastal parameter.

• Boundary level fluctuations. Fluctuations in the thermocline and the halocline boundary layers may be very important for surface and deep water retention times, especially in deep and open coasts.

• Local winds may create a water exchange in all coastal areas, especially in comparatively small and shallow coasts.

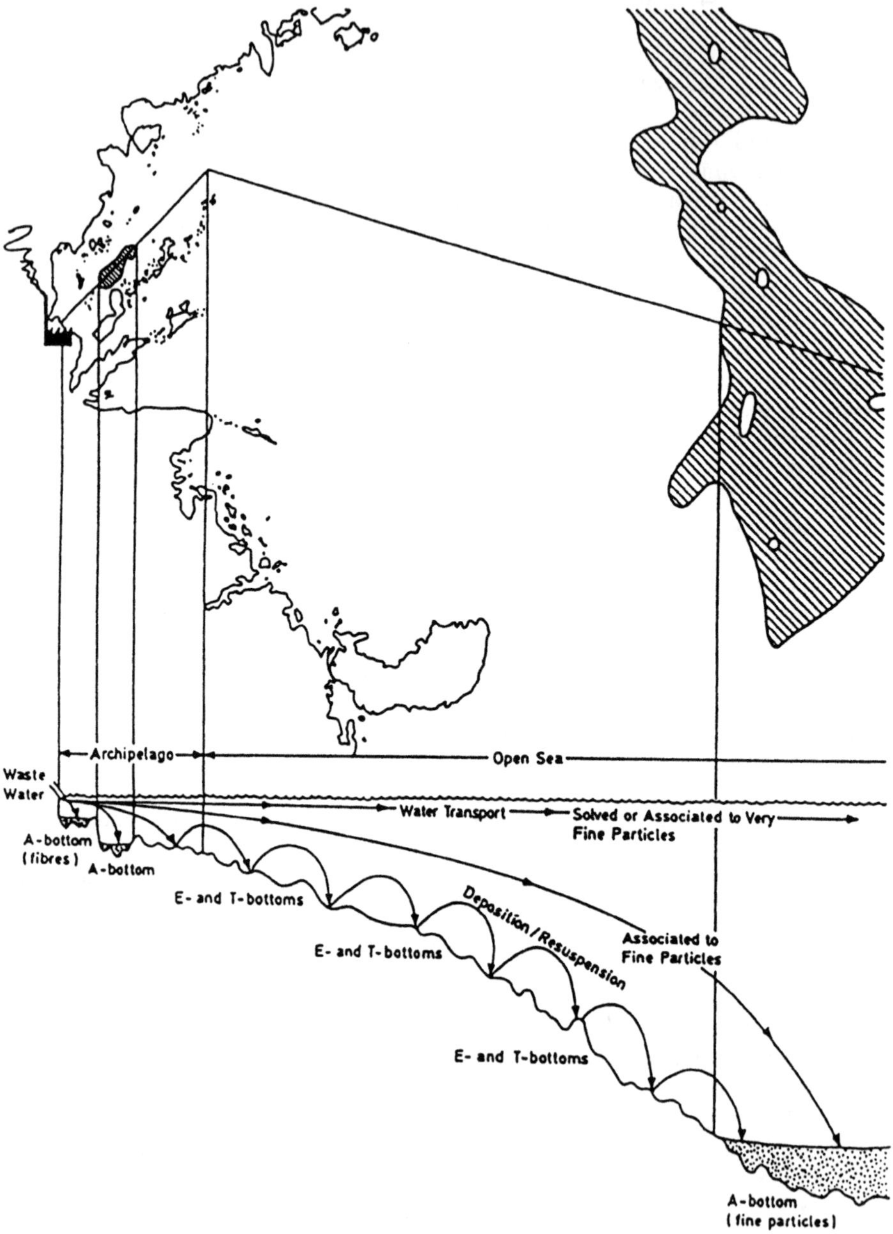
Archipelago
Open Sea
Waste Water
Water Transport
Solved or Associated to Very Fine Particles
A-bottom (fibres)
A-bottom
E- and T-bottoms
Deposition / Resuspension
E- and T-bottoms
Associated to Fine Particles
E- and T-bottoms
A-bottom (fine particles)

Fig. 2.7. Bottom dynamic conditions outside Norrsundet in the Bothnian Sea. The point-source emissions are distributed in many cycles. The accumulation areas in the open water part of the Bothnian Sea appear at water depths below 50 m. Sediment sample from shallower areas, often transportation areas, may vary very much, and may be very old. Modified from Håkanson et al. (1988b).

• Thermal effects. Heating and cooling, e.g., during warm summer days and nights, may give rise to water level fluctuations which may cause a water exchange. This is especially true in shallow coasts since water level variations in such areas are more linked to temperature alterations in the air than is the case in open water areas.

• Coastal currents (see Figure 2.9) are large, often geographically concentrated, shore-parallel movements in the sea close to the coast. They may have an impact on the water retention time, especially in coasts with a great topographical openness.

In theory, it may be possible to distinguish driving processes from mixing processes. In practice, however, this is often impossible. Surface water mixing causes a change in boundary conditions, which causes water exchange, and so on.

The theoretical water retention time (T) for a coastal area is the time it would take to fill a coast of volume V if the water input from rivers is given by Q and the net water input from the sea by R, i.e.: $T = V/(Q + R)$. This definition does not account for the fact that actual water exchange normally varies temporally, areally and vertically, e.g., above and beneath the thermocline. In the following model, Ts is the theoretical surface water retention time and Td the theoretical deep water retention time. These concepts evidently only apply to stratified coastal areas (see Figure 2.10).

There are several methods of determining or estimating the water exchange (see Håkanson et al., 1984, 1986).

• The fresh water input to a bay may be used as a "tracer", and the salinity or the conductivity of the water may be used to calculate the water exchange by means of mass-balance equations. The conductivity is then measured by CTD-sonds (CTD stands for measurement of Conductivity, Temperature and Depth). Many kinds of instruments are commercially available. If the CTD-sonds are used to determine the water exchange in an estuary, one must measure salinity or conductivity inside and outside the given coast, as well as the coastal volume and the input of fresh water via tributaries.

• Instead of using the fresh water as a "tracer", one may also use real tracers, like dye tracers (e.g., rhodamine, a red dye). The dye tracer method requires quite a lot of special equipment and trained staff (see Figure 2.11).

• The direction and velocity of water currents may be measured quite simply with inexpensive current meters, which automatically measure the mean direction of the flow and water velocity for the period of registration. If several current meters are placed in a given section, the water exchange can be determined for the coastal area as a whole.

• The water level may be measured by different types of gauges (permanent or mobile, continuously recording or manually handled). The water exchange can be determined from gauge data (differences in water level over time) if the area and volume of the coast are known.

One fundamental abiotic factor that, together with the morphometry (i.e., the size and shape of the coastal area), sets the framework for the biological life is the salinity (Figure 2.3). The salinity in the open water areas outside the coastal zone varies from

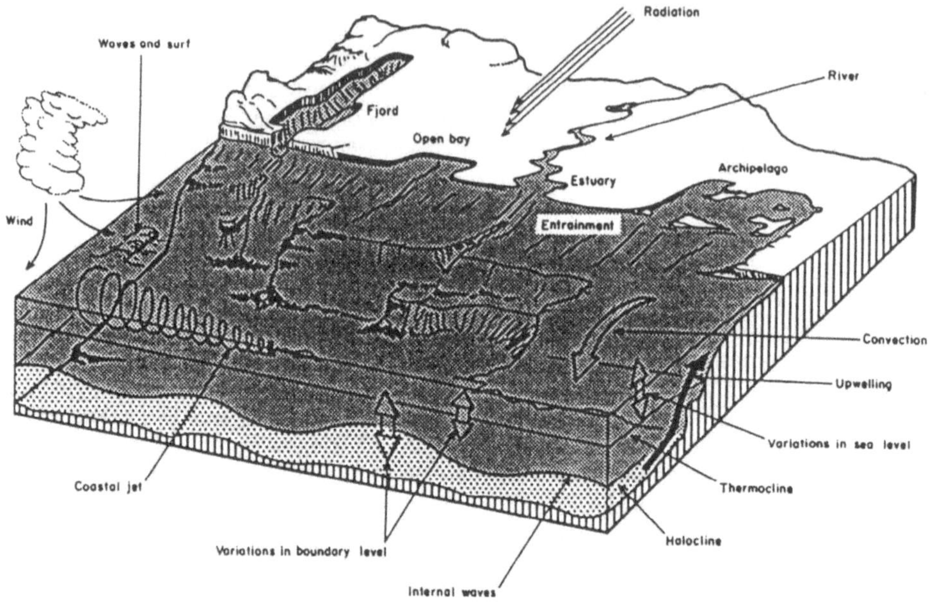

Fig. 2.8. Schematic illustration of key processes regulating water exchange in coastal areas (modified from Håkanson et al., 1986).

about 2-4‰ in the Bothnian Bay, to 4-6‰ in the Bothnian Sea, to 6-8‰ in the Baltic Proper, and to values in the range 20-30‰ in the Kattegat and Skagerrak. The deep water at any site is generally more saline than the surface water. Between 50 and 70 m there is a fairly rapid increase in salinity. This steep salinity gradient is called the halocline. Beneath the halocline, one finds saltier, denser water with a salinity in the range 10-12‰.

Outside the coastal zone, the water down to a depth of about 20 m is significantly warmer (11-12°C) in September than in January (when the temperature at this layer is about 2°C in the Baltic). When the water is thermally stratified during summer, there is often a zone with a steep gradient in temperature. This gradient is called the thermocline, the water above the thermocline is often referred to as the surface water, and the water beneath the thermocline is called the deep water or the bottom water. This means that in late summer, one finds warmer, less saline water on top of colder water with approximately the same salinity. During winter, the temperature increases steadily from about 2°C at the surface to about 4°C at the bottom.

These two boundary layers, the thermocline and the halocline, are acting as "bottoms" or "glide surfaces" for the transport of water and pollutants carried by the water.

In the Baltic coastal zone, the maximum thermocline is generally at a water depth of about 10 m (Persson et al., 1994; see Figure 2.10 for illustration). The theoretical deep water retention time is generally longer than that of the surface water (this also depends on the volume of the deep water), and the deep water is often exchanged episodically. The mixing between surface water and deep water is generally small during stratified conditions, but efficient during homothermal conditions (see Persson and Håkanson, 1996).

2.2.3. FUNDAMENTAL SEDIMENTOLOGICAL PRINCIPLES AND PROCESSES FOR COASTAL AREAS

The sediments reflect what is happening in the water mass and on the bottom — they may be regarded as a tape-recorder of the historical development and are often called "the geological archive". The sediments also affect the conditions in the water via, e.g., resuspension processes animals living in the sediments; they both play a fundamental role in the ecosystem. By extracting sediment cores and conducting a number of analyses, information is obtained on changes that have taken place in the ecosystem (see Jonsson, 1992).

The grain size and/or the composition of the material are often used as criteria to distinguish different sediment types. Alternatively, one can differentiate between different sediment types by means of functional criteria (like erosion, transportation and accumulation) of coarse sediments (friction material) or fine sediments (cohesive material). In geo-ecological contexts, it is common to focus on the finer materials most easily set in motion/resuspension and having the highest capacity to bind pollutants.

Geochemically, fine sediments behave differently as compared to coarse materials. From the basic Stokes's equation for settling particles (see Figure 2.12), as well as for convenience, the limit between coarse and fine materials can be set at a particle size of fine sand (0.06 cm).

The generally hard or sandy sediments within the areas of erosion and transport (ET) often have low water content, low organic content, and low concentrations of nutrients and pollutants (see Table 2.1). The conditions within the T-areas are, for natural reasons, variable, especially for the most mobile substances, like phosphorus, manganese and iron, which may react rapidly to alterations in the chemical "climate" (given by the redox potential) of the sediments. Fine materials may be deposited for long periods during calm weather conditions. In connection with a storm or a mass movement on a slope, this material may be resuspended and transported up and away, generally in the direction towards the A-areas in the deeper parts, where continuous deposition occurs. Thus, resuspension is a most natural phenomenon in T-areas.

It should be stressed that fine materials are rarely deposited as a result of simple vertical settling in natural aquatic environments. The horizontal velocity component is generally at least 10 times larger, sometimes up to 10,000 times larger, than the vertical component for fine materials or flocs which settle according to Stokes's law (see Bloesch and Burns, 1980; Bloesch and Uehlinger, 1986).

Figure 2.13 gives a schematic summary of central concepts and processes in coastal sedimentology. For literature on general sedimentological processes in coastal areas, see, e.g., Muir Wood, 1969; Stanley and Swift, 1976; Dyer, 1979; McCave, 1981;

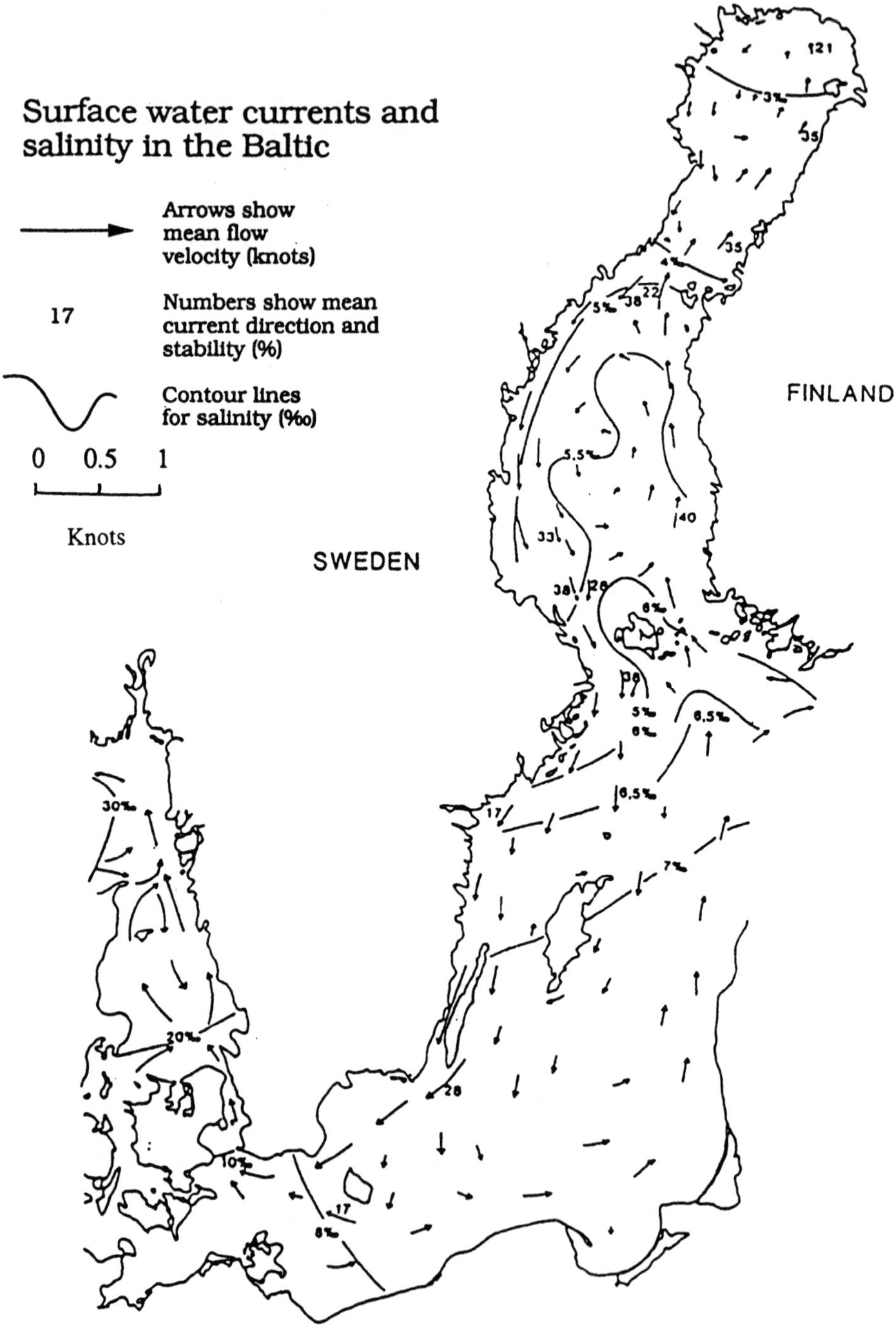
Surface water currents and salinity in the Baltic
Arrows show mean flow velocity (knots)
17
Numbers show mean current direction and stability (%)
Contour lines for salinity (‰)
0 0.5 1
Knots
SWEDEN
FINLAND

Fig. 2.9. Surface water currents and salinity in the Baltic. Modified from FRP (1978).

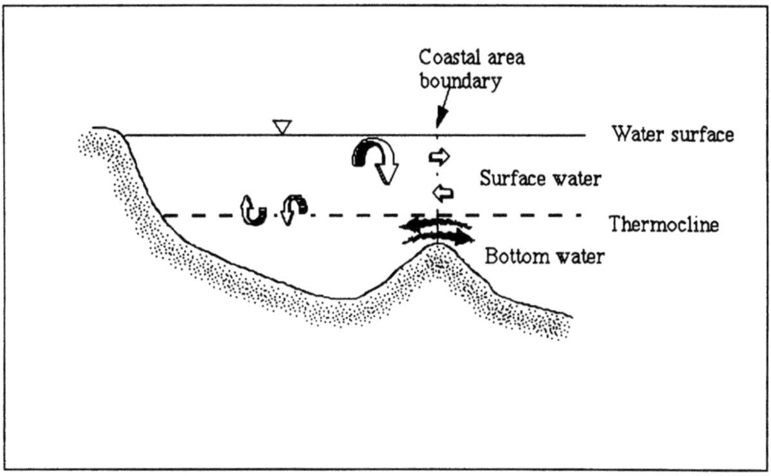

Fig. 2.10. Illustration of thermal stratification of a coastal area (modified from Carlsson et al., 1999).

Seibold and Berger, 1982; Postma, 1982; Håkanson, 1986. From this figure, one should note the following points.

1. **River action** dominates the sedimentological properties in estuaries, where deltas may be formed if the amount of sandy materials carried by the tributaries is large enough. Generally, the rate of sedimentation and the grain size decrease logarithmically with the trajectory distance from the mouth of the river.

2. **Wind/wave action** generally dominates the bottom dynamic conditions in coastal areas (more than 90% of the coastal areas of the Baltic are dominated by these processes). The rate of sedimentation normally increases from the wave base and with increasing water depth below the wave base, a process often referred to as "sediment focusing". The coarsest materials (sand, gravel, etc.) are often found in shallow waters.

3. **Current action** (unidirectional flows) can dominate in certain coastal areas. Then the "Hjulström-curve" gives the relationship between critical erosion and critical deposition of materials.

Slope-induced (gravity) turbidity currents appear on bottoms inclining more than about 4-5% (Håkanson, 1977), and bioturbation generally prevails in oxic sediments, where the macro- and meiofauna cause a mixing of the sediments. Human activities, like trawling, can also influence the sediment resuspension (Floderus, 1989).

2.2.4. COASTAL MORPHOMETRY

There are great differences concerning topographical, hydrodynamical and biological characteristics of different coastal regions. Figure 2.14 gives a differentiation of different coastal types for the Baltic. It should be noted that archipelagos dominate this coastal landscape.

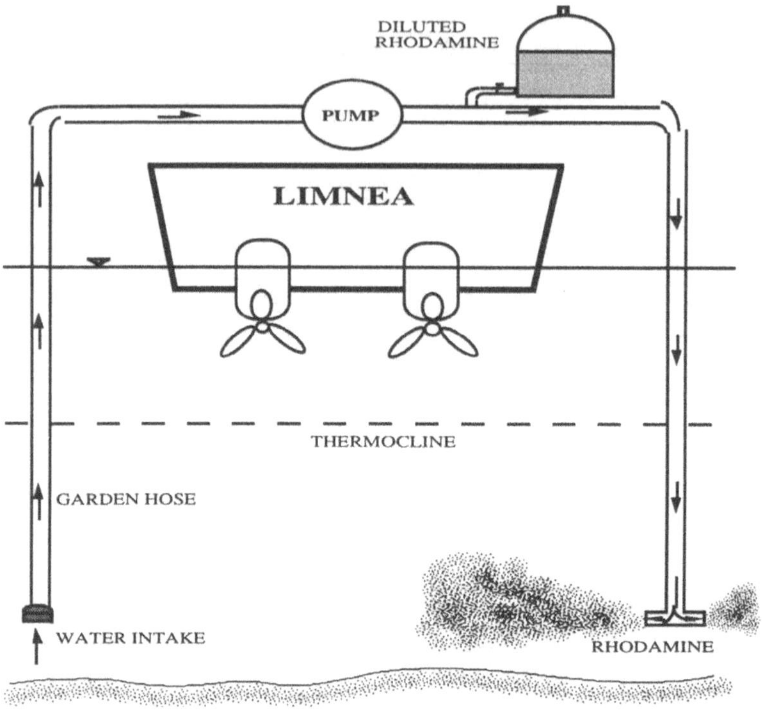

Fig. 2.11. Illustration of dye injection below the thermocline. Limnea is the name of the research vessel (modified from Persson and Håkanson, 1991).

What the coast looks like, i.e., the topographical character of the coast, determines to a great extent how the water system functions ecologically (Håkanson, 1999). Thus, it is important to define and measure certain key morphometrical parameters that may be used for quantitatively describing and classifying coastal areas.

Figure 2.14 gives a genetic (descriptive) coastal classification, which presents a large-scale qualitative picture of the distribution of various coast types. It does not give any hard data for calculations of the water exchange and/or the bottom dynamic conditions. Coastal classifications, like Figure 2.14, provide important background information as to why a given area looks the way it does, and may also be used for general discussions on the suitability/unsuitability of various coasts for different purposes.

Morphometrical parameters may be classified into the following three categories (Håkanson et al., 1984; see Figure 2.15).

• Size parameters.

The size parameters maximum depth (Dmax, m), water surface area (Area, km^2), total area (TotA, km^2), area of all islands (Ai, km^2) and water volume (V or Vol, km^3) are all easy to define (Pilesjö et al., 1991) and need no further explanation.

The section area (At, km^2; see Figure 2.16 for definition) is defined as the contact surface between the coastal area and the surrounding sea or adjacent coastal

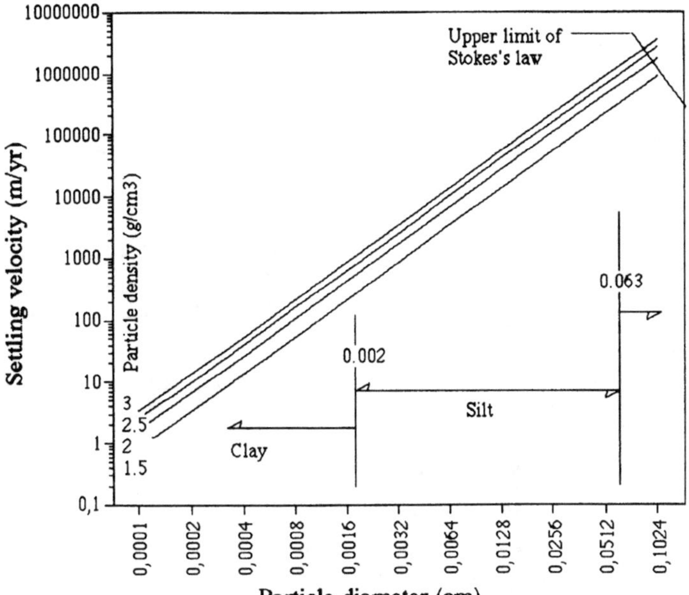

Fig. 2.12. The relationship between the settling velocity (v of spherical particles) in water, particle diameter and particle density (typical values for humus is ≈ 1.5, clay ≈ 2.4, quartz ≈ 2.7, dolomite ≈ 2.8) as given by Stokes's law (at 20°C).

Table 2.1. Typical values of different sediment parameters from the Swedish coastal zone in different bottom dynamic conditions (modified from Håkanson et al.. 1984).

	Erosion	Transportation	Accumulation
PHYSICAL PARAMETERS			
Water content (% ww)	< 50	50 - 75	> 75
Organic content (loss on ignition, % dw)	< 4	4 - 10	> 10
NUTRIENTS (mg/g dw)			
Nitrogen	< 2	10 - 30	> 5
Phosphorus	0.3 - 1	0.3 - 1-5	> 1
Carbon	< 20	20 - 50	> 50
METALS			
Iron (mg/g dw)	< 10	10 - 30	> 20
Manganese (mg/g dw)	< 0.2	0.2 - 0.7	0.1 - 0.7
Zinc (μg/g dw)	< 50	50 - 200	> 200
Chromium (μg/g dw)	< 25	25 - 50	> 50
Lead (μg/g dw)	< 20	20 - 30	> 30
Copper (μg/g dw)	< 15	15 - 30	> 30
Cadmium (μg/g dw)	< 0.5	0.5 - 11.5	> 1.5
Mercury (ng/g dw)	< 50	50 - 250	> 250

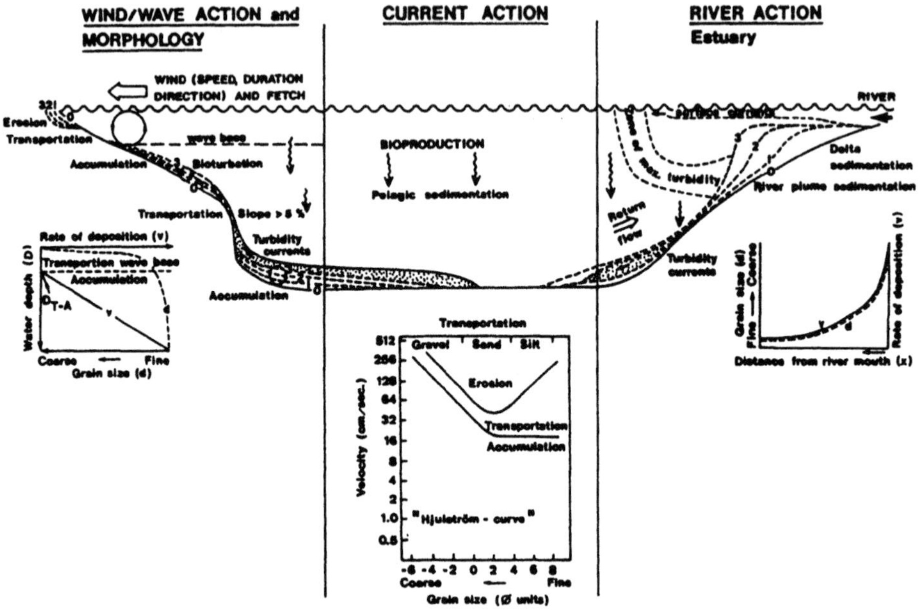

Fig. 2.13. Illustration of major concepts and processes in coastal sedimentology (modified from Håkanson et al., 1984).

areas. If the coastal area has more than one opening/contact surface with the surrounding sea, the section area is the sum of these surfaces.

Area and At are used in the definition of the Exposure (Ex; see Figure 2.16), which is used to estimate the theoretical surface water retention time (Ts). Both of these are of great importance in predicting radiocesium concentrations in water, sediments and biota in coastal areas. This is exemplified in Figure 2.17, which gives results from a sensitivity analysis for a coastal area. Since the coastal area (Area) and the section area (At) are directly related to the exposure (Ex = 100·At/Area) and hence to theoretical surface water retention time (Ts), it is easy to understand why it is important to use relevant and reliable data for these two parameters in this model for radiocesium. The default value for the coastal area (Area = A = 5 km^2; see Table 2.3 later) has been multiplied by 0.1 and 10 to illustrate how this would influence radiocesium concentration in water and in 1000 g ww piscivores. The larger the coastal area, the greater the direct fallout (in Bq/m^2). The figures illustrate that it is important to have access to a realistic value for the coastal area for the peak predictions in water and the predictions of the Cs-concentrations in fish since they influence two of the most crucial factors regulating the abiotic and biotic Cs-concentrations in the coastal area, i.e., direct fallout and the Ts-value.

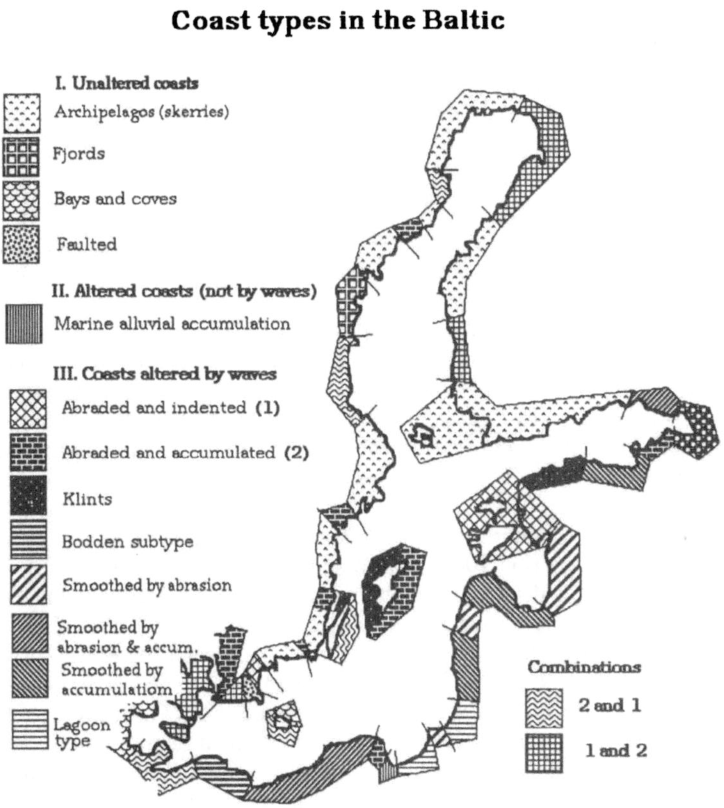

Fig. 2.14. Coast types in the Baltic. Redrawn from Voipio (1981).

Note that the size of the coastal area influences many, sometimes contrasting, processes:

(1) the larger the coastal area, the greater the direct fallout, and hence the higher the potential Cs-concentrations in water and biota;

(2) the larger the coastal area, the larger the volume, and hence the lower the Cs-concentrations in water and biota;

(3) the larger the coastal area, the larger the potential ET-areas and the larger the internal loading, the higher the Cs-concentration in water and several species of fish.

This means that it is often very difficult, and sometimes impossible, to give logical, clear-cut mechanistic reasons as to how a certain x-variable, like coastal area, would influence a certain target y-variable, like Cs-concentrations in water. This demonstrates the important and even unique role of mathematical models in the contexts of water management and research dealing with transport, biouptake and the effects of

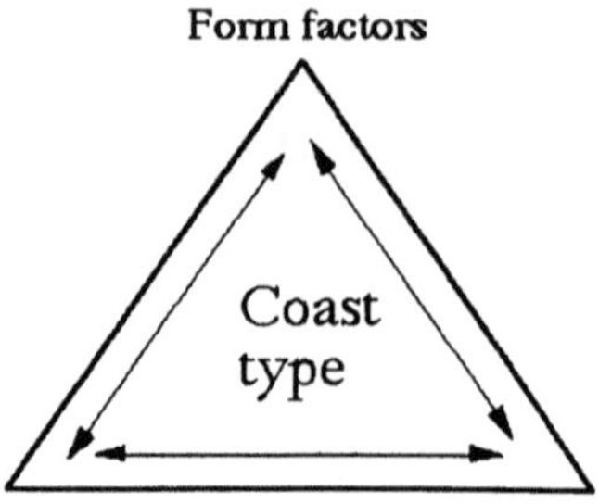

Fig. 2.15. A given coastal area may be described by different morphometrical parameters, which can be categorised according to size, form and special parameters (after Håkanson and Rosenberg, 1985).

water pollutants in complex aquatic ecosystems. Quantitative models are powerful tools for sorting and ranking the importance of various processes and for disclosing inter-dependencies of this kind. The result given in Figure 2.17 is a good example of this. One should note that, from the presuppositions given for this model, the Cs-concentrations in 1-kg piscivores are probably highest in the largest coastal area (while all else is constant), lowest in the in-between (default) coastal area and in-between in the smallest coastal area.

Figure 2.18 gives a similar sensitivity analysis for section area. The default value for the section area (a section 1000 m wide and 10 m deep, i.e., At = 10,000 m^2) has been multiplied by 0.1 and 10 to illustrate how this would influence radiocesium concentrations in water and 1-kg piscivores. The larger the section area, the greater the exposure, the faster the surface water turnover time, the greater the impact from the sea, and hence the greater the dilution if the Cs-concentrations in the sea are lower than the Cs-concentrations in the coastal water, or the opposite if the Cs-concentrations in the sea are higher. The figures illustrate that under the defined default conditions, the Cs-concentrations in water during the time just after the fallout depend very much on the value used for the section area (the initial dilution), but neither the "tail" values of the curve nor the Cs-concentrations in piscivores depend very much on the value used for At. The main reason for this is that under default conditions, the Cs-concentrations in the sea decline quite slowly. This may be valid for semi-enclosed seas like the Baltic, but not for Mediterranean or Atlantic coastal areas (which will be discussed later on).

• Form parameters, generally determined from size parameters.

The mean depth (Dm, m) is simply the volume (Vol in m^3) divided by the area (in m^2). The mean slope (xm in %) can be defined in many ways (see Håkanson, 1981; Pilesjö et al., 1991). A rather comprehensive formula is:

$$xm = (lo + 2 \cdot L) \cdot Dmax/20 \cdot n \cdot Area \quad (2.1)$$

where

lo	= the length of the shoreline (in m); preferably normalised to account for differences in map scale at different calculations (see Håkanson, 1981);
L	= the total length of contour lines (in m)
Dmax	= the maximum depth (m);
n	= the number of contour lines (= isolines);
Area	= coastal area (m^2).

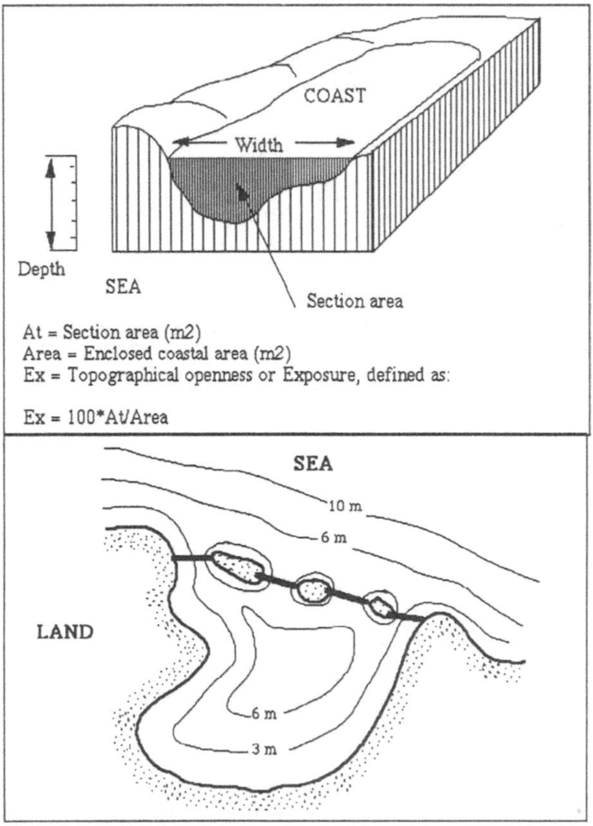

Fig. 2.16. Illustration and definition of the section area (At) and the topographical openness or exposure (Ex), which is used to define a coastal area. The borderlines toward the open sea or adjacent coastal areas are drawn where the exposure attains a minimum value (modified from Håkanson, 1998).

This means that two more morphometric parameters, lo and L, as well as the number of contour lines (n) are needed to determine the mean slope. It is easy to determine these driving variables from GIS-maps, but if such digital bathymetric information is not available, the mean slope can be predicted quite well ($r^2 = 0.78$ based on data from 39 Baltic coastal areas; data from Wallin et al., 1992) from the following regression equation. This equation has been derived for this model from the presupposition that no more driving variables should be accepted to predict the mean slope in addition to those which are already used to predict the theoretical deep water retention time. Thus, xm can be predicted from the following equation:

$$xm = 4.59 + 0.19 \cdot Dm - 1.86 \cdot (At/Vol)^{0.5} + 5.7 \cdot \log(Drel) \quad (2.2)$$

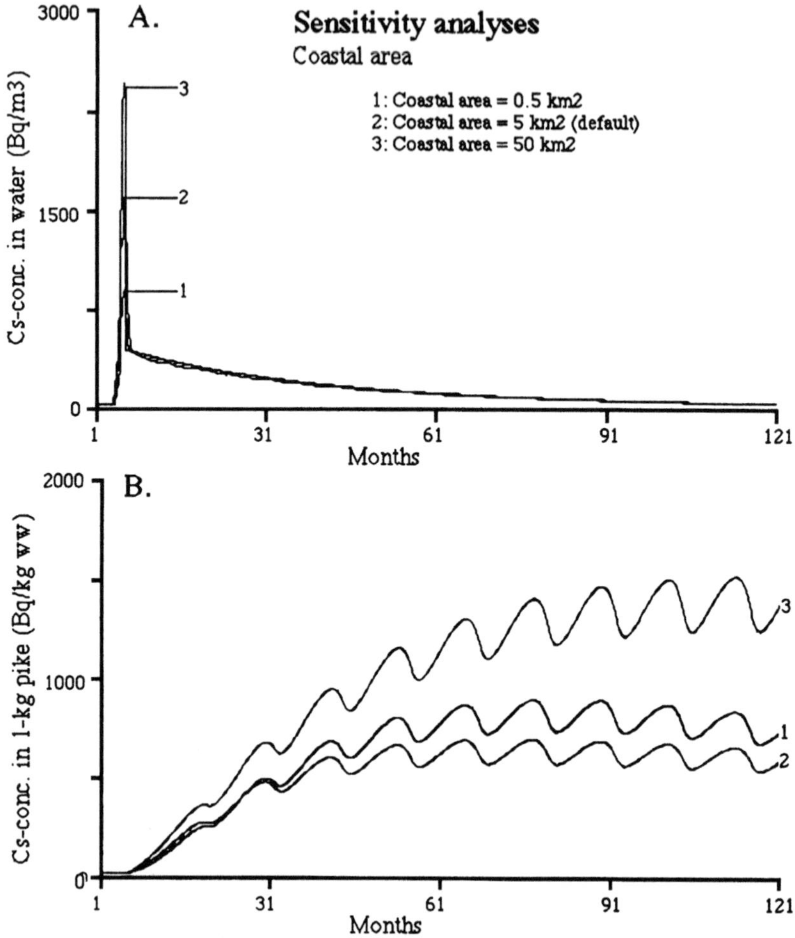

Fig. 2.17. Sensitivity analysis for coastal area variations. The default value for the coastal area (5 km^2) has been multiplied by 0.1 and 10 to illustrate how this would influence radiocesium concentrations in water (A) and in 1-kg piscivores (like pike) (B).

where

Drel = the relative depth; see Figure 2.19 for definition;

At = the section area in km^2; note that At is given in km^2 and not m^2 in this equation for xm;

Vol = the coastal volume (= Dm·Area) in km^3; note that Vol is given in km^3 and not m^3 in this equation for xm;

Dm = the mean depth (= Vol/Area) in m.

The relative depth (Drel, %) is defined as the ratio between the maximum depth and the mean diameter of the coastal area. The relative depth gives a measure of the stability and stratification of water masses (Eberly, 1964). Small and deep coastal areas have high Drel-values while large and shallow coastal areas have low Drel-values.

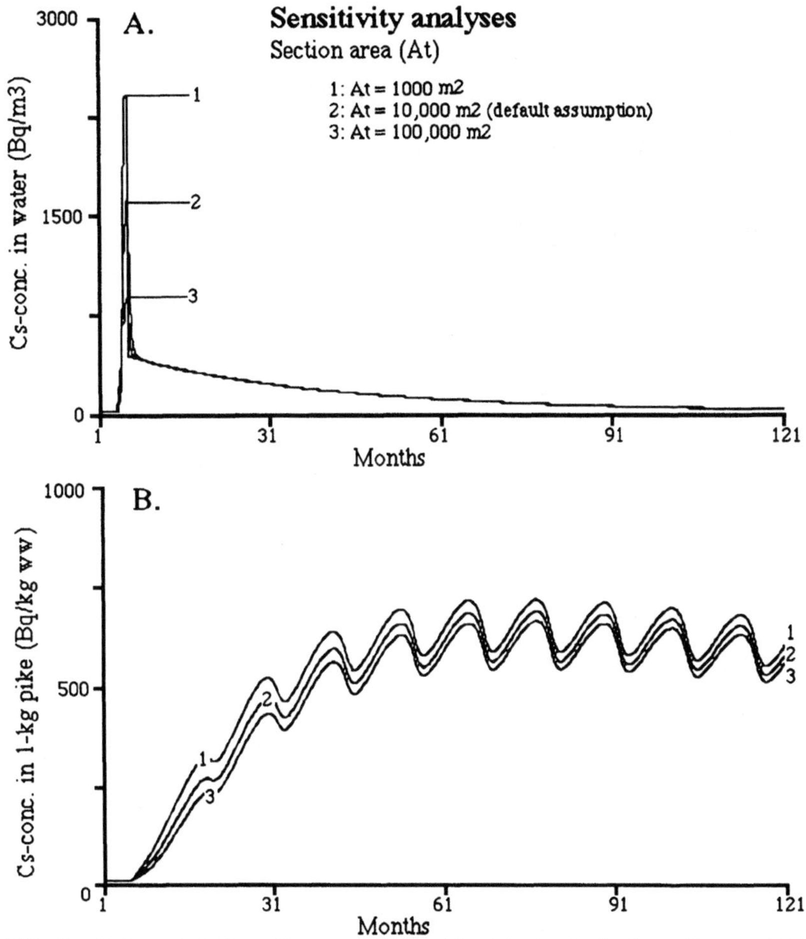

Fig. 2.18. Sensitivity analysis for section area variations. The default value for the section area (10,000 m^2) has been multiplied by 0.1 and 10 to illustrate how this would influence radiocesium concentrations in water (A) and in 1-kg piscivores (like pike) (B).

The volume development (or form factor, Vd, dim. less) is defined as the ratio between the water volume and the volume of a cone with a base equal to the water surface area and with a high equal to the maximum depth (see Håkanson, 1981), i.e.:

$$Vd = (a \cdot Dm \cdot 1000) / (1/3 \cdot Dmax \cdot 1000 \cdot a) = 3 \cdot Dm/Dmax \qquad (2.3)$$

where Dm is the mean depth (m) and Dmax is the maximum depth (m).

The Vd-value is used in several contexts in the model for radiocesium.

Figure 2.20 gives a sensitivity analysis for one of the most important form parameters, the mean depth. The mean depth (Dm) influences several processes in the model, e.g., the settling rate for particulate radiocesium and the volume and hence

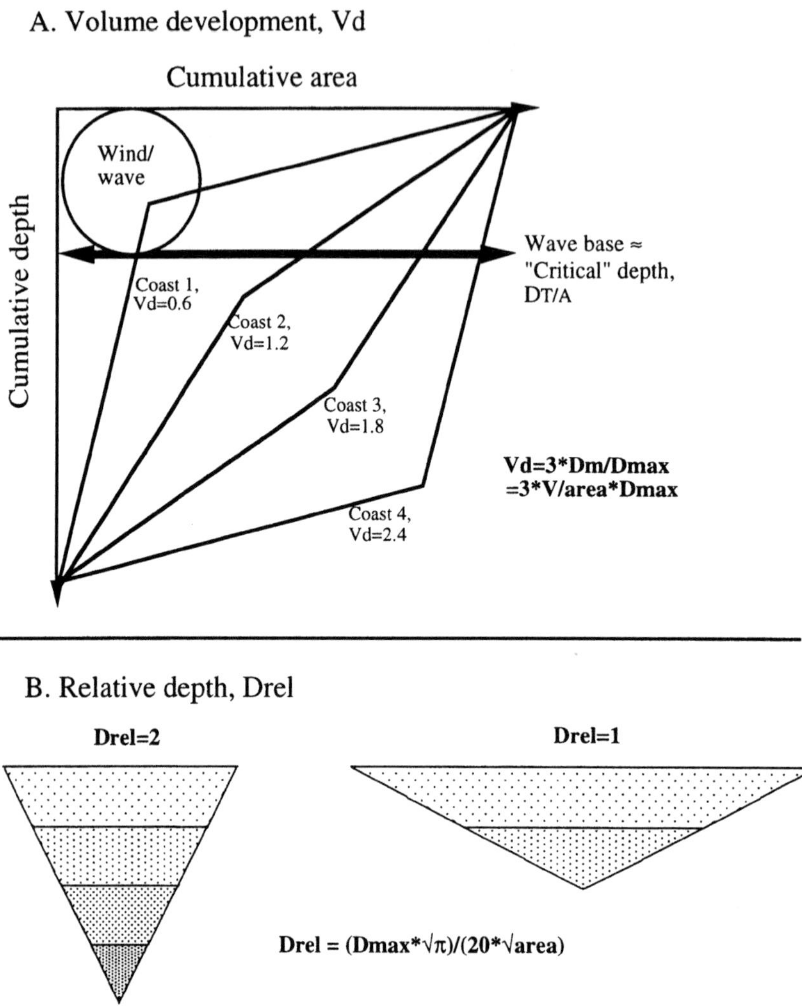

Fig. 2.19. Top. Schematic illustration of relative hypsographic curves for four coastal areas with different form factors (volume development, Vd = 0.6, 1.2, 1.8, or 1.4). Areas with small Vd-values have large shallow areas subject to wind- and wave induced resuspension. Areas with high Vd-values have greater volumes which may increase autochthonous sedimentation and steeper slopes where slope-induced resuspension may occur.
Bottom. Illustration of the relative depth (Drel) and the concept of sediment focusing, i.e., that the deposition of material increases with depth.

everything related to volume, such as Cs-concentrations in water, and hence Cs-concentrations in biota. The default value for the mean depth is 9 m. The mean depth has been varied in geometric steps from 1.5 m to 24 m in this test and the results are given for the radiocesium concentrations in water (A) and in 1-kg piscivores (like pike) (B). It is very important to have access to reliable data on Dm, since the mean depth

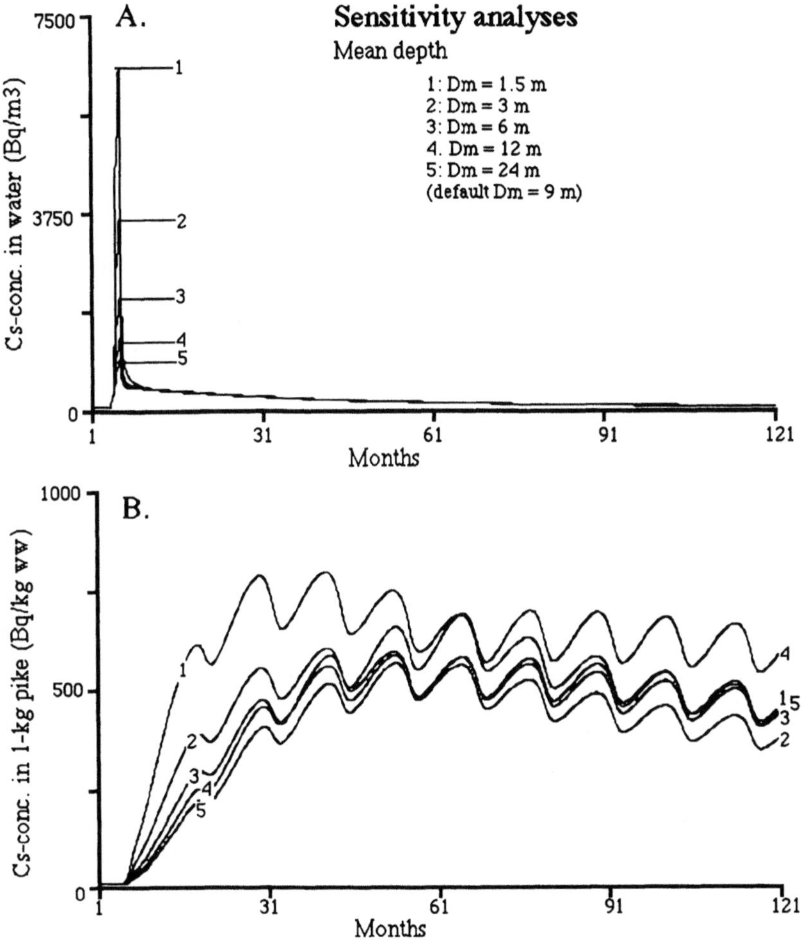

Fig. 2.20. Sensitivity analysis for the mean depth. The mean depth has been varied in geometric steps from 1.5 m to 24 m in this test and the results are given for radiocesium concentrations in water (A) and in 1-kg piscivores (like pike) (B).

influences many of the key factors regulating the response to a given fallout. Many of these factors work in opposite directions, e.g., a large Dm means a greater volume and, hence, a lower Cs-concentration in water and also a smaller portion of the area above the wave base, and less resuspension. Also, there is a more efficient entrapment of particulate radiocesium, which means that more cesium will be retained in the coastal area, and hence there will be higher possible Cs-concentrations in benthivores and piscivores. Only mathematical models can sort out the relative roles of such complicated processes. These sensitivity tests for the default coastal area indicate that the deeper coastal areas are likely to have fish with relatively lower Cs-concentrations than those in shallow coastal areas. However, coastal areas with "in-between" mean depths may have higher

Cs-concentrations, as exemplified by curve 4 in Figure 2.20, which illustrates that in the long run the areas with mean depths of 12 m (in this scenario) have the highest Cs-concentrations in 1-kg piscivores, whereas the fish in the shallowest coastal areas (curve 1) have the highest Cs-concentrations during the initial phase.

• Special parameters.

The exposure (Ex in %) is, as pointed out, defined as the ratio between the section area (At) and the coastal area (Area) (see Figure 2.16). Large, enclosed and deep coastal areas have low Ex-values while smaller, more exposed and shallower coastal areas have higher Ex-values.

The filter factor (Ff, km^3) is a measure quantifying the amount of wave energy from the sea that reaches a coastal area after passage through the surrounding archipelago (from Pilesjö et al., 1991). Ff is basically a modification of the effective fetch (see Håkanson et al., 1984), which gives the wind/wave impact at one single point. The filter factor, on the other hand, accounts for the wind/wave impact on the entire coastal area. The filter factor is defined as:

$$Ff = \Sigma(\Sigma(\cos(a_i)\cdot x_i\ )\cdot At)_j \qquad (2.4)$$

where a_i and x_i are angles in radians and distances to land in km, respectively, for a imaginary group of radials starting from the mid-point of one of the opening sections of the coastal area (At) (see Figure 2.21).

The index, i, ranges from 1 to 315, which means that the angle varies between $-\pi/2$ and $\pi/2$ radians in relation to the normal of the opening section. The sum of $\cos(a_i)\cdot x_i$ for the 315 angles is multiplied by the section area of the opening (At, see above). This product is calculated for all the openings in the coastal area (j), and the sum of them is the filter factor.

A large section area and an open surrounding archipelago results in a high Ff-value. It should be noted that the filter factor cannot be calculated if one or more radials is/are endless. However, the influence of the effective fetch (x_i) on the waves and the bottom dynamic conditions is more or less constant if the fetch exceeds 100 km (Håkanson et al., 1984), so the maximum length of a radial has been set to 100 km in the calculation of the filter factor.

The mean filter factor (MFf) is defined as the ratio between the filter factor (Ff) and the number of openings (i.e., section areas) toward the sea and/or adjacent coastal areas. It is used in the sub-model to predict the theoretical deep water retention time (Td).

Figure 2.22 summarises the most frequently morphometrical parameters in this model. From this background, we will now focus on the coastal model for radiocesium.

2.3. The Model

An important conceptual basis for the derivation of this model is the following statement: "The topographical (= morphometrical) character of the coastal area regulates the way the coastal area functions". This means that coastal morphometry is a very important structural component for this model. Morphometric parameters are easily accessed from standard bathymetric charts from modern GIS-systems; much

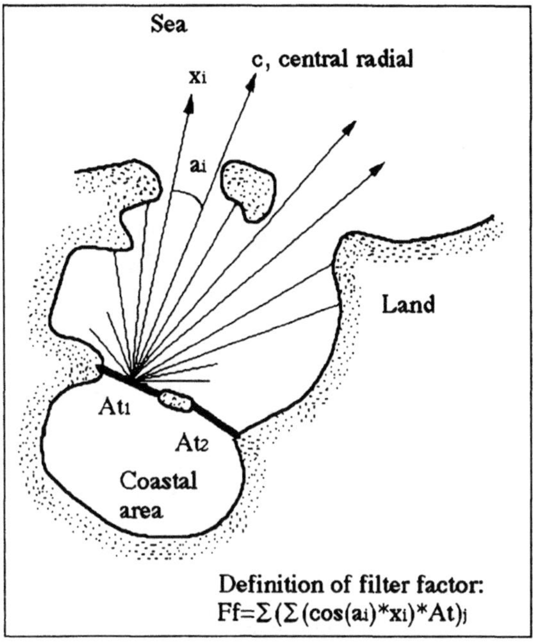

Fig. 2.21. Illustration and definition of the filter factor (Ff).

information can (see Pilesjö et al., 1991) be derived from such sources. This will be demonstrated in the following sections.

2.3.1. PRESUPPOSITIONS

There are several prerequisites for this model:

1. In this modelling approach, it is important to define the fundamental unit, i.e., the coastal ecosystem. This is not a diffuse food-web structure, but a defined geographical area. An ecosystem is a rather uniform entity with respect to its defining characteristics, like Cs-concentrations in water and biota. One very important question concerns the definition of coastal ecosystems. The question is where to place the boundaries with the sea and/or adjacent coastal areas. It is crucial to use a technique that provides an ecologically meaningful and practically useful definition of the coastal ecosystem. How should one define this area so that, e.g., parameters, like mean depth, filter factor and relative depth can be relevant as model variables (x) for predicting the target variables?

Arbitrary borderlines can be drawn in many ways and the morphometric parameters of such areas would be devoid of meaning in relation to target variables. The approach in this work (from Håkanson et al., 1984 and Pilesjö et al., 1991) assumes that the borderlines are drawn at the topographical bottle-necks so that the exposure (Ex) of the coast from winds and waves from the open sea is minimised. It is easy to use the Ex-value as a tool to test different alternative borderlines and define the coastal ecosystem where the Ex-value is minimal. It is a prerequisite for this model that the

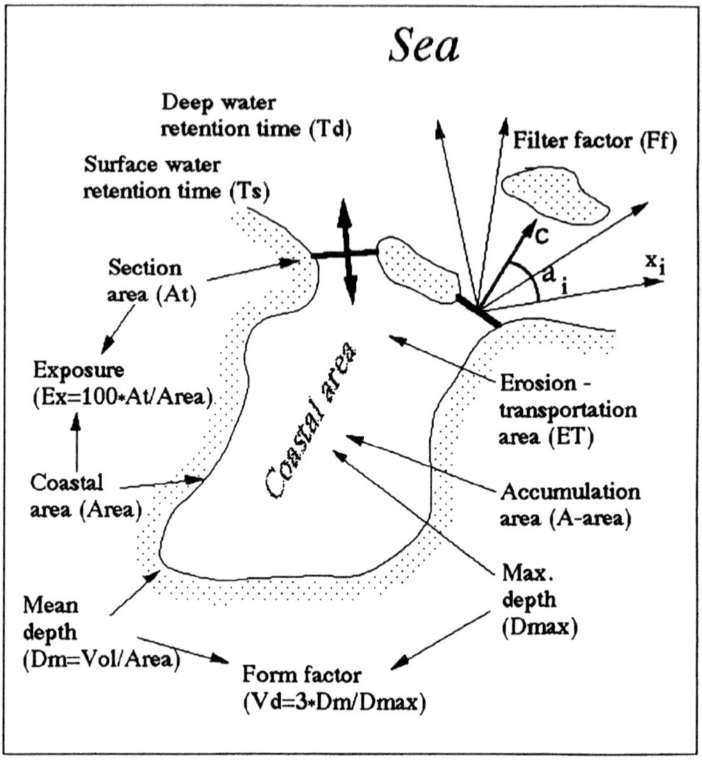

Fig. 2.22. Illustration of some important morphometrical parameters used in the model for radiocesium in coastal areas.

coastal area is defined according to this method, which is easy to apply using GIS-systems.

It is always important to define the presuppositions of any model. When and where will it apply? The definition of the ecosystem boundaries is one crucial aspect of this for the model.

Since most scientists working in coastal areas do not have this ecosystem perspective, this method is not well known. By presenting this approach here, it is hoped that the method will become better known, so that more scientists can see the benefits of applying the ecosystem perspective to coastal studies.

2. It is very costly and laborious to empirically determine the theoretical (or characteristic) surface water retention time (Ts), which is necessary if concentrations of water pollutants are to be predicted along with ecosystem effects related to increases in concentrations. We have discussed several field methods for determining Ts, and they are all costly. Ts is certainty one of the major regulating factors for inflow, outflow and retention of substances in coastal areas. However, Ts, which is defined from the net flow of water from the sea and adjacent coastal areas and the volume of the surface water, depends on many more or less stochastic processes. This makes it difficult to give a reliable prediction of Ts at a given time. In any coastal area Ts may be

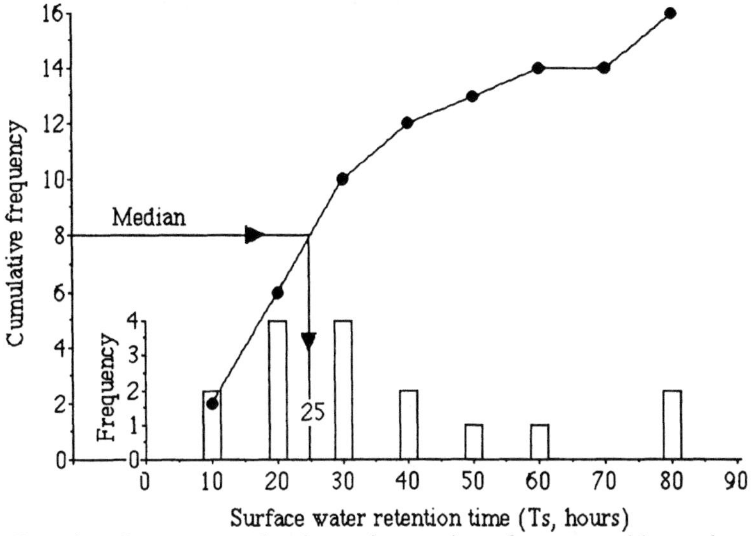

Fig. 2.23. Illustration of measurement data from a dye experiment from a coastal bay to determine the theoretical surface water retention time (Ts). Modified from Håkanson et al. (1984).

indefinitely long on a calm summer day and very short (say 1 hour) in connection with a storm or a sudden change in air pressure. Ts always emanates from a frequency distribution (see Figure 2.23). This model assumes that the characteristic Ts-value, the median value (25 hours as illustrated in Figure 2.23), is used. There is an evident uncertainty associated with such Ts-values. We will perform sensitivity and uncertainty tests to see how important this uncertainty is relative to other uncertainties for the target variables in this model (Cs-concentrations in water and fish). As a rule of thumb, one can say that the costs of establishing a frequency distribution such as that in Figure 2.23 from traditional field measurements (using dye, current meters, etc.) is about 20,000 USD for one coastal area. It is not very meaningful to build a management model if it is a prerequisite that such field work first must be carried out to determine Ts as a driving variable. This means that it is of major importance that Ts can, in fact, be predicted very easily from one coastal morphometric variable, i.e., exposure (see Figure 2.24).

$$Ts = e^{(3.49-4.33\cdot\sqrt{Ex})}/30 \qquad (2.5)$$

This model for radiocesium uses predicted Ts-values (in months) such as that given by eq. 2.5. It has been shown (Persson et al., 1994) that Ts can be predicted very well ($r^2 = 0.93$) with this morphometric regression based on the exposure (Ex), which is a function of section area (At) and coastal area (Area) (see Figure 2.16). The range of this model for Ts is given by the minimum and maximum values for Ex in Table 2.2. That is, Ex should vary from 0.002 to 1.3, and the model should not be used when the tidal range is > 20 cm/day or for estuaries, where the fresh water discharge must also be accounted for. For open coasts, i.e., when Ex > 1.3 and/or the mean filter factor (MFf) > 8, Ts is calculated not by this equation but from a model based on coastal currents

Sub-models for surface and deep water retention times and ET-areas

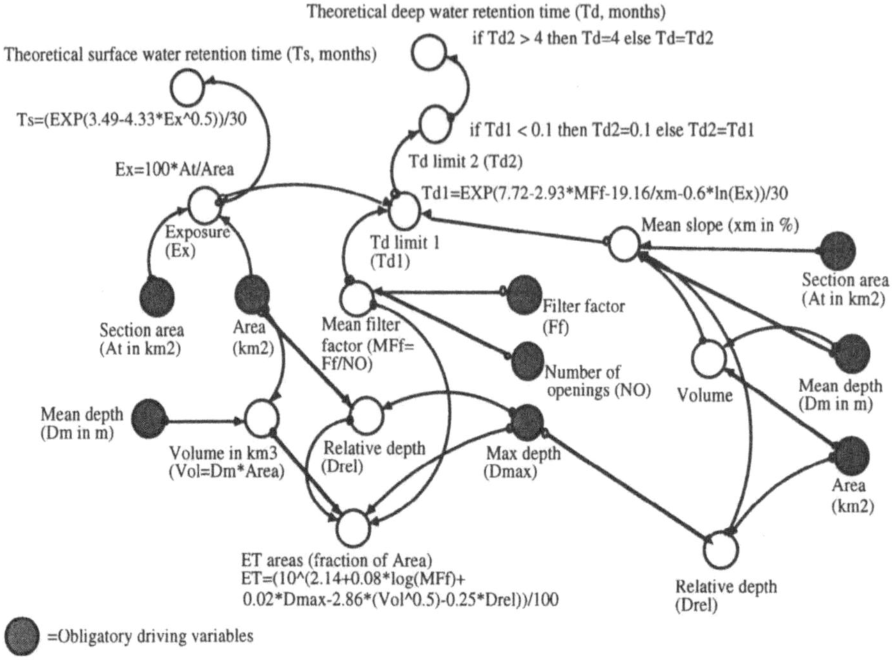

Morphometric equations:
xm=4.59 + 0.19*Dm - 1.86*((At*1000)/Vol)^0.5 + 5.7*log(Drel)
Drel=Dm*(π^0.5)/(20*((Area)^0.5))

Fig. 2.24. The morphometric sub-models for surface (Ts) and deep water (Td) retention times and ET-areas.

(the u-formula, see eq. 2.23) and/or from the tidal range formula (the dH-formula, see eq. 2.24).

3. The same arguments can be given for the deep water retention time (Td). It is very difficult and costly to determine Td with, e.g., the dye method illustrated in Figure 2.11. The costs per area are even higher for Td than for Ts, but Td can also be predicted from readily available morphometric parameters (r^2 = 0.82) see Figure 2.24. This empirical model is based on three morphometric parameters (for further information, see Persson and Håkanson, 1996), which can all be determined easily from digital bathymetric maps.

a. The mean filter factor (MFf); step 1; r^2 = 0.39. This is the filter factor (see Figure 2.21) divided by the number of openings; MFf is the most important morphometric factor for the Td-variability among the coastal areas studied; the more open the coastal area outside the given coastal area, the smaller the size of the "wind/wave-energy filter", the larger the MFf-value and the shorter the theoretical deep water retention time. This is certainty mechanistically reasonable.

Table 2.2. Typical values for key morphometric parameters and statistical features for the Baltic coastal areas. lo = shoreline length (km). From Håkanson (1994), based on data from 39 coastal areas.

Size parameters		Units	Mean	Min.	Max.
Dmax	maximum depth	m	19.9	3.4	46.9
Area	water surface area	km^2	4.7	0.87	13.9
At	section area	km^2	0.01	0.00049	0.082
Vol	water volume	km^3	0.03	0.0002	0.18
Form parameters					
Dm	mean depth = V/Area	m	6.6	0.93	13.8
xm	mean slope	%	4.2	0.79	8.2
Drel	relative depth = $Dmax \cdot \sqrt{\pi}/20 \cdot \sqrt{Area}$	%	0.97	0.20	2.7
F	shore irregularity = $lo/(2 \cdot \sqrt{(\pi \cdot Area)}$	%	187	104	507
Vd	form factor = 3•Dm/Dmax	-	1.01	0.45	1.47
Special parameters					
Ex	exposure = 100•At/Area	-	0.35	0.002	1.3
Ff	filter factor	km^3	5.5	0.06	30.7
MFf	mean filter factor	km^3	1.5	0.01	8.0
ET	ET-areas	%	79.4	19.1	100

b. Mean slope (xm); step 2; $r^2 = 0.78$. The the second most important morphometric factor for Td is xm: the larger the slope per volume unit, the deeper the coast and the longer the theoretical deep water retention time. This is also logical.

c. Exposure (Ex; see Figure 2.16 for definition); step 3; $r^2 = 0.82$. Ex is the third most important morphometric factor: the more open the coast, the larger the section area (At) relative to the coastal area (Area), and the shorter the theoretical deep water retention time.

The range of this model for Td is given by the minimum and maximum values for the three model variables in Table 2.2. This Td-formula is, evidently, only applicable to thermally stratified coastal areas not influenced by tides and or fresh water fluxes since it is based on data from Baltic coastal areas. If Td cannot be estimated with this formula, the default assumption is to set $Td = 10 \cdot Ts$ for thermally stratified coastal areas and $Td = 100 \cdot Ts$ for coastal areas with more or less permanent stratification (like deep fjords).

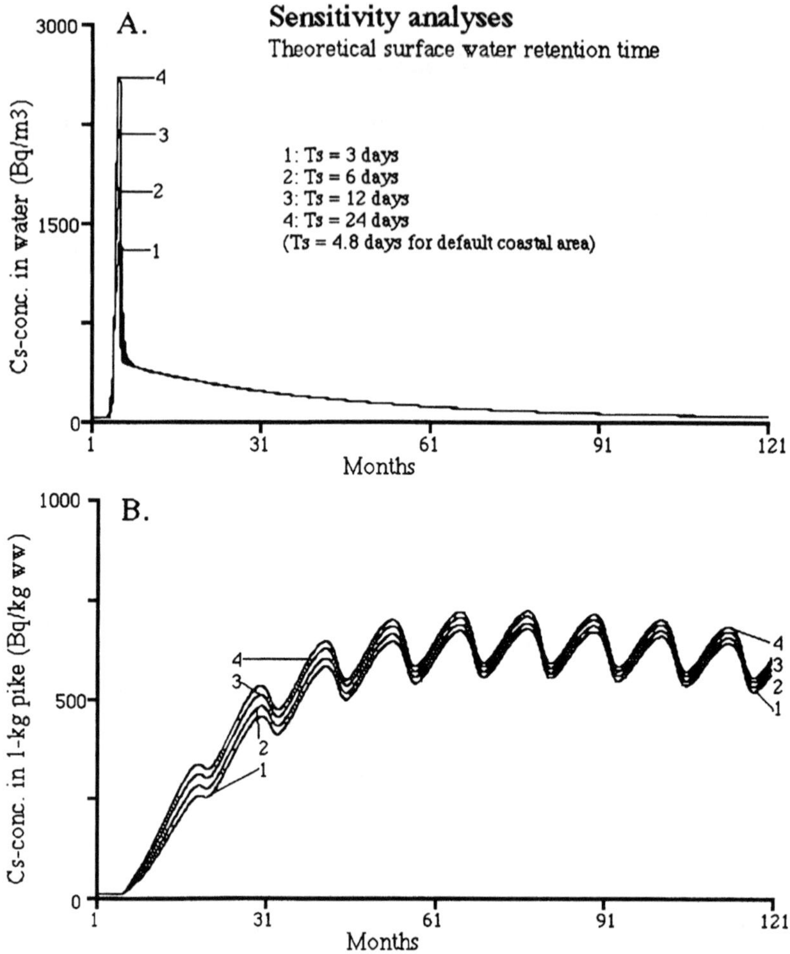

Fig. 2.25. Sensitivity tests for different values for Ts, the surface water retention time (3, 6, 12 and 24 days) in relation to radiocesium concentration in water (A) and in 1-kg piscivores (pike) (B).

all cases given in Figure 2.25, however, is rather efficient and the differences in Cs-concentrations in water are relatively small after the second month after the fallout. The predicted Cs-concentrations in piscivores (and especially benthivores) depend on the amounts of radiocesium retained in the sediments, and there are no major differences among the coastal areas in this respect in the cases given in Figure 2.25. However, the results depend very much on the assumptions concerning the inflow of radiocesium from the sea. If less radiocesium is transported from the sea to the coast, then the predicted Cs-concentrations in biota will depend more on the selected Ts-value.

The conditions in the default coastal area are given in Table 2.3, and Figure 2.26 gives a simulation for Cs-concentrations in water (curve 1), 1-kg piscivores (curve 2), 10 g ww planktivores (curve 3) and active A-sediments. These results are used as reference values in many of the following tests and they are given here to facilitate those interpretations.

Table 2.3. A compilation of data used to describe the default conditions of the coastal area.

Coastal area = $5 \cdot 10^6$ m^2	Catchment area = 10·Coastal area
C_{sea} = see text	Coastal current = 2.5 cm/sec
Default Kd = 800000 (l/kg)	Default settling velocity = 10 m/month
D_{OAL} = 0.5 (dim . less)	Filter factor = 3 km^3
Inorganic nitrogen conc. (IN) = 20 μg/l	K-concentration = 1 mg/l
Loss on ignition = 7% dw	Max. depth = 30 m
Mean annual precipitation = 750 mm	Mean depth = 9 m
Month of fallout (May) = 5	Outflow areas (OA) = 0.2 (dim. less)
Resuspension rate = 1/21 (1/days)	Salinity = 6‰
Secchi depth = 5 m	Section area (At) = 10,000 m^2
Soil permeability factor = 25 (dim. less)	Susp. part. matter = 3 mg/l
Thickness of active layer = 0.05 m	Tidal range = 2 cm/day
Water content = 70% ww	Mean slope = 4%
Number of openings (section areas) = 2	

Figure 2.27 gives results for sensitivity tests for Td, the deep water retention time. Note that the morphometric Td-formula assumes that there exists a thermal stratification and periods of mixing; a similar approach could also be used for coasts with a chemocline, e.g., a halocline. Here, different Td-values have been tested in relation to radiocesium concentrations in water and 1-kg piscivores (pike). Under default conditions, the model predicts a Td-value of just 0.1 months, since the deep water volume is relatively small in the default area. One would expect Td-values to vary in the range from Td ≈ Ts to a Td-value of about 4 months (for coastal areas which mix in the spring and the autumn) to Td ≈ 100·Ts in stratified fjords. If such Td-values are used, these simulations (Figure 2.27) indicate that Td has no major influence on the predictions of Cs-concentrations in water and biota. Thus, one can often accept a rather crude estimate of the Td-value. It is logical that the predictions are not so sensitive to errors and uncertainties in rates like 1/Td, which generally regulate minor and not major Cs-fluxes. The most important conclusion that can be drawn from the tests in Figures 2.25 and 2.27 is that it is generally more important to have reliable data on Ts than on Td. This means that it is essential to use methods that can provide as reliable data as possible for Ts, and that these methods should be driven by readily available driving variables.

4. The same arguments can also be given for the bottom dynamic conditions, i.e., whether ET- and/or A-areas dominate in the given coastal area. The sediment type and the bottom dynamic conditions can be determined with different methods, like echo-sounding, sediment sampling and sediment analysis (see Håkanson and Jansson, 1983). All those methods require field work and trained personnel and they are all costly and time-consuming. This means that it is an advantage that the fractions of ET-areas (ET) and A-areas (A = 1-ET) can, in fact, be predicted quite well with empirical models based on readily available morphometric parameters (see Håkanson, 1986; Persson and Håkanson, 1995). This model for radiocesium uses a new sub-model to predict ET. It has been developed for this radiocesium model and from a requirement that additional driving variables other than those already included in the model should not be permitted.

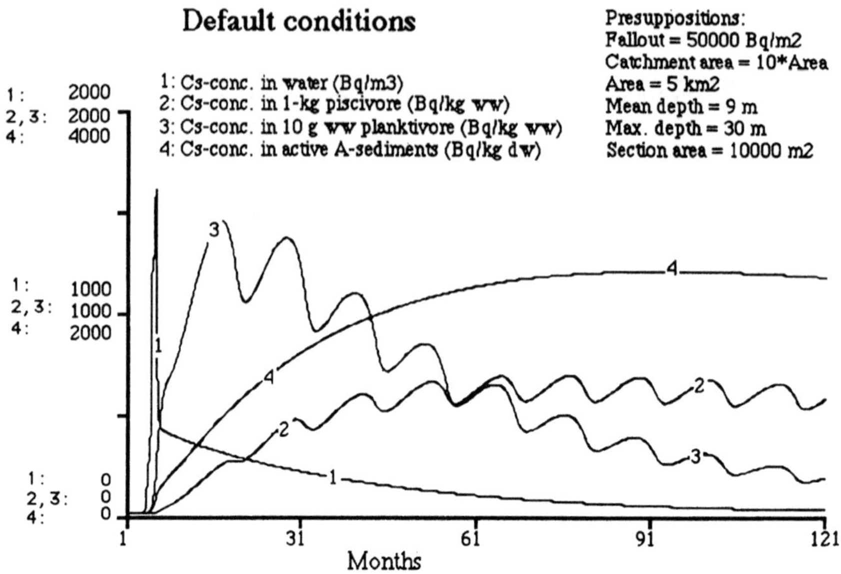

Fig. 2.26. Default conditions. This figure gives predictions concerning radiocesium concentrations in water, 10 g ww planktivores, 1000 g ww piscivores and A-sediments (0-5 cm) under default conditions, i.e., fallout 50,000 Bq/m^2 during month 5 (Chernobyl situation), coastal area 5 km^2, section area 10,000 m^2, mean depth 9 m, maximum depth 30 m, salinity 6 ‰ and with the catchment area sub-model included. The figure illustrates the very fast and efficient dilution concerning Cs-concentrations in water and the response concerning the Cs-concentrations in biota and active A-sediments.

This means that the models presented by Persson et al. (1995), which provide somewhat higher r^2-values that the model given here because they use driving variables like the mean depth of the coastal area beneath the wave base and criteria to determine whether the given coastal areas does or does not have a direct contact with the sea, will not be used.

The driving variables for this model are given below (see Table 2.4).

Step 1; coastal volume (Vol in km^3); r^2 = 0.28; the larger the volume, the larger the percentage of A-areas and the smaller the ET-areas.

Step 2; mean filter factor (MFf); r^2 = 0.41; this is certainly a logical model variable; the larger the MFf-value, the more open the coast outside the given coastal area (= the small the wave "energy filter"), the greater the wind/wave impact and the larger the ET-areas.

Step 3; the maximum depth (Dmax); r^2 = 0.47; the greater the maximum depth, the greater the percentage of slope-induced ET-areas.

Step 4; the relative depth (Drel); r^2 = 0.65; this is a morphometric parameter that is known to be important in the contexts of sediment focusing and bottom dynamic conditions (see Håkanson and Peters, 1995); the greater the Drel-value, the greater the percentage of slope-induced ET-areas.

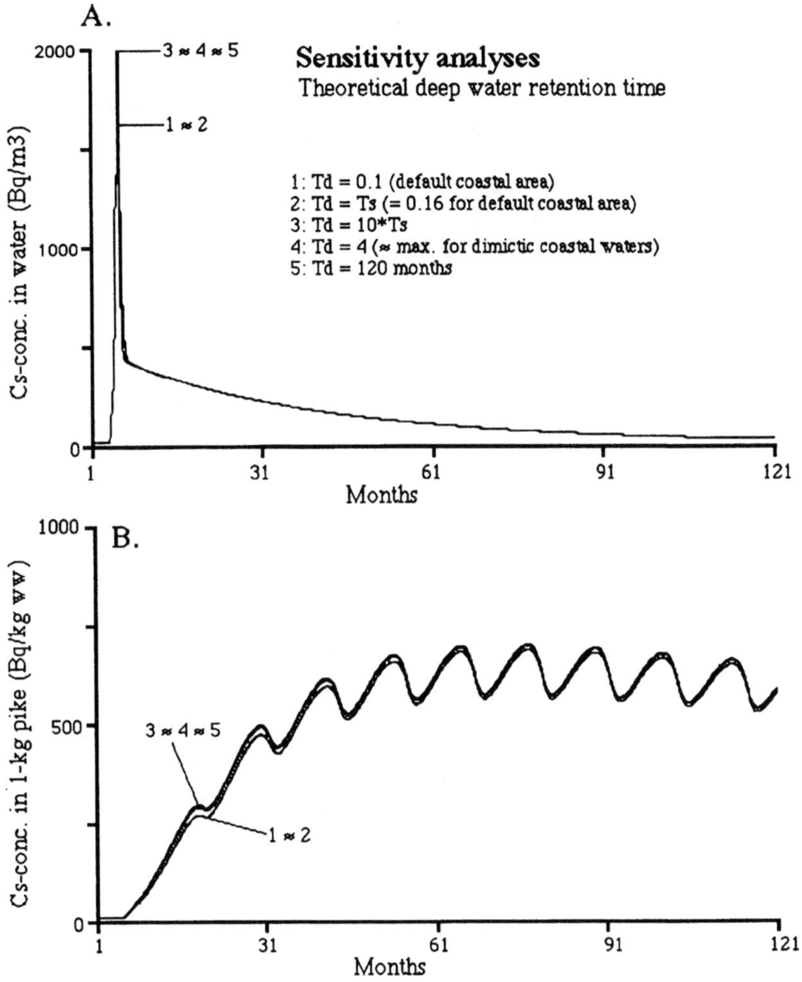

Fig. 2.27. Sensitivity tests for different approaches for Td in relation to radiocesium concentration in water (A) and in 1-kg piscivores (pike) (B).

The maximum depth and the relative depth are, by definition, related to one another, but they are not very highly correlated ($r^2 = 0.41$ for 39 Baltic coastal areas) because they represent different form elements of coastal areas.

The range of this model for ET is given by the minimum and maximum values for the four model variables in Table 2.2. This formula is based on data from Baltic coastal areas and this sets the domain for this formula. If ET cannot be estimated with this morphometrical regression equation, the default assumption in this model for radiocesium is to set ET to 0.7, i.e., as 70% of the coastal areas.

Figure 2.28 gives a sensitivity analysis for the influence of bottom dynamic conditions (ET-areas) on the radiocesium concentrations in 10 g ww planktivores (A) and 100 g ww benthivores (B), which are likely to be more dependent on the bottom dyna-

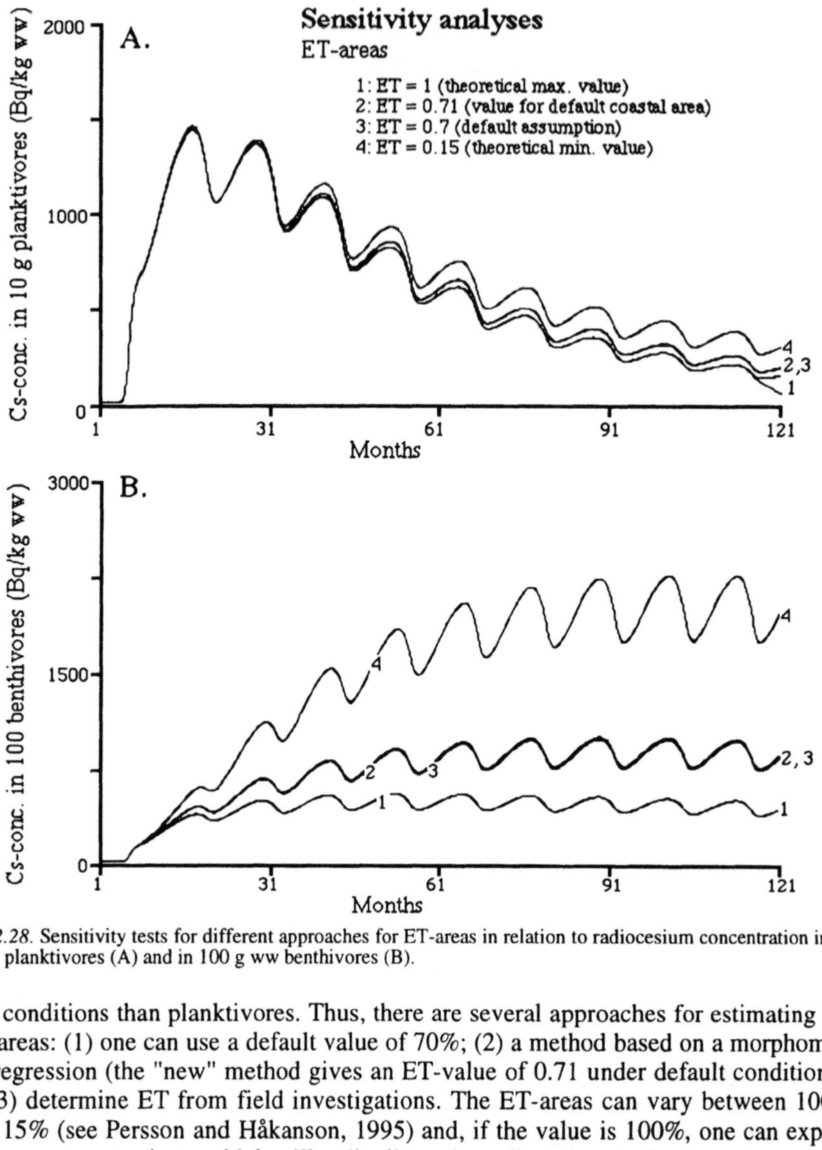

Fig. 2.28. Sensitivity tests for different approaches for ET-areas in relation to radiocesium concentration in 10 g ww planktivores (A) and in 100 g ww benthivores (B).

mic conditions than planktivores. Thus, there are several approaches for estimating the ET-areas: (1) one can use a default value of 70%; (2) a method based on a morphometric regression (the "new" method gives an ET-value of 0.71 under default conditions); or (3) determine ET from field investigations. The ET-areas can vary between 100% and 15% (see Persson and Håkanson, 1995) and, if the value is 100%, one can expect frequent resuspensions, which will redistribute the radiocesium in the coastal area. The figures illustrate that under default conditions the model predicts that it is not very important for Cs-concentrations for planktivores to have reliable data on ET-areas. It is, however, much more important for the Cs-concentrations for the benthivores. The radiocesium concentrations in water and biota depend mainly on the total amount in the coastal area and this is primarily influenced by the inflow and outflow processes rather than the internal processes. This means that sub-models related to other sediment processes, like sediment Kd, sediment redox potential, etc. may be disregarded in this

Table 2.4. The "ladder" for the empirical model for predicting the fraction of ET-areas. These results are based on data presented by Persson and Håkanson (1996) and methods presented by Håkanson and Peters (1995) for 38 Baltic coastal areas. Note that this empirical model may produce uncertain prediction if it is used for other coastal areas, like areas influenced by heavy tides. y = log(ET); ET in % of Area. Vol = volume in km^3; MFf = mean filter factor; Dmax = max. depth in m; Drel = relative depth. F ≥ 4.

Step	Model variable	r^2-value	Model
1	√Vol	0.28	y=3.03-0.98·x1
2	log(MFf)	0.41	y=2.07-1.13·x1+0.08·x2
3	Dmax	0.47	y=2.02-1.49·x1+0.08·x2+0.01·x3
4	Drel	0.65	y=2.14-2.86·x1+0.08·x2+0.02·x3-0.25·x4

coastal model. It is more crucial to have access to proper values for sedimentation than for the internal fluxes that depend on sedimentation. To illustrate this point, the following sensitivity analyses focus on sedimentation and the factors regulating this process.

The following two simulations are meant to illustrate a situation in which it is important to properly account for internal loading. Figure 2.29 first gives a compilation and ranking of all abiotic monthly fluxes (Bq/month) for the default coastal area. One should note that there are MAJOR differences for the given processes. The flow from the sea and the outflow from the coast totally dominate the fluxes. Sedimentation on ET-areas, resuspension from ET-areas to water, secondary transport to the coast from outflow areas of the catchment (OA to coast), secondary transport to the coast from inflow areas of the catchment (IA to coast), transport from active to passive A-sediments and sedimentation on A-areas (Sed A) are less important fluxes, and diffusion from A-areas is the smallest flux of all. This means that in this coastal area one could omit diffusion and the predicted Cs-concentrations in water and fish, i.e., the two target variables in this model, would be almost the same. However, the ranking of the fluxes is different in different coastal areas and depends very much on the presuppositions for the inflow from the sea, so diffusion can be more important in other situations, and even in this coastal area under different presuppositions.

This is shown in Figure 2.30. This scenario is meant to illustrate a situation in which all the radiocesium is being dumped directly on the A-sediments. In such cases, the secondary transport from land to water and the inflow from the sea would be negligible. In the case given in Figure 2.30, all the radiocesium (10^{10} Bq) was added to the active A-sediments in month 10. Under these conditions, diffusion is one of the MAIN fluxes, almost equal to outflow from the coastal area, but still smaller than the transport from active to passive A-sediments.

Thus, in this definition of the coastal area, the morphometric regression equations for Ts, Td and ET are important prerequisites for this model for radiocesium.

2.3.2. EQUATIONS

2.3.2.1. The catchment sub-model - tributary inflow to estuaries

The catchment sub-model is the same as that already presented in section 1 (Figure 1.4). This sub-model handles the inflow to the coast via the tributaries. The direct deposition (fallout multiplied by area) is the net sum of dry and wet fallout onto the catchment.

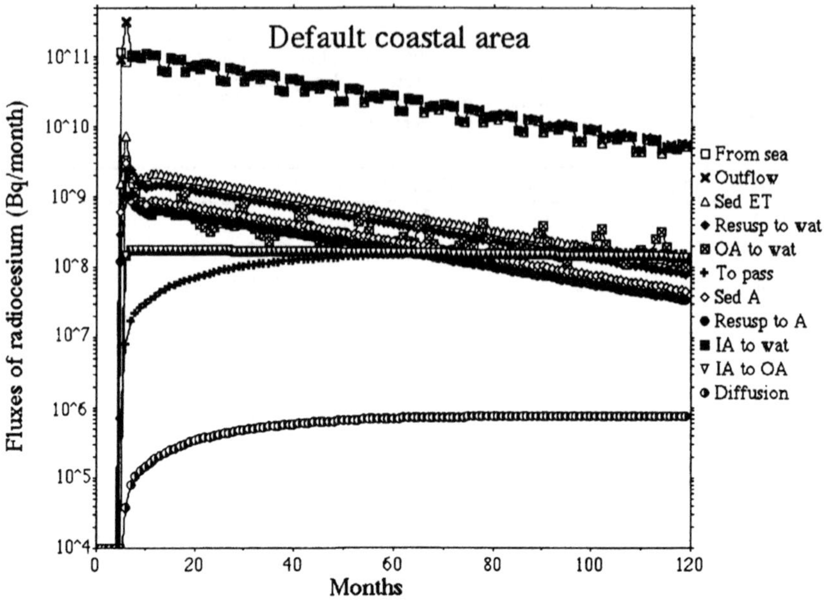

Fig. 2.29. This figure gives all fluxes in the model to and from the default coastal area in Bq/month. The ranking of the fluxes is: transport from the sea > outflow from the coast > sedimentation on ET-areas > resuspension from ET-areas to water > secondary transport from outflow areas to water > transport from active to passive A-sediments > secondary transport from inflow areas to outflow areas = secondary transport from inflow areas directly to water > sedimentation on A-areas > resuspension from ET to A-areas > diffusion.

The tributary load is a function of the amount of ^{137}Cs in the catchment multiplied by an outflow (transfer or retention) rate.

Figure 2.31 gives a sensitivity analysis for the catchment area sub-model. This test illustrates the predictions with and without the catchment area sub-model under otherwise default conditions. The default value for the size of the catchment area is 50 km^2 (or 10·Area, Area = the coastal area; default value 5 km^2). Curve 3 illustrates the results when the catchment is 500 km^2. The figure shows that this would influence the Cs-concentration (Bq/kg ww) in 1-kg pike and the total Cs-concentration in the water (Bq/l) relatively little. The larger the catchment, the greater the impact from the secondary load, and hence the higher the concentrations in water and biota. Note, however, that the influence on the Cs-concentration in water is also quite small when the catchment area is large. The main reason for this is that there is a very fast water exchange between the coast and the sea, which means that the conditions in the coast are dominated by the fast water exchange between the sea rather than by tributary input. "The conditions in the coast are governed by the conditions in the sea."

The month of the fallout is important, not so much for the Cs-concentrations in water, but more for the Cs-concentrations in biota; see Figure 2.32. The critical point is when fallout happens relative to (1) ice-conditions, (2) seasonal characteristics of the tributary water discharge and (3) seasonal variation in the bioproduction. These conditions vary among and within coastal areas, e.g., with latitude. For the predicted Cs-concentrations in water in the default coastal area, the value selected for the month of the fallout will primarily move the peak of the curve. For the Cs-concentrations in biota, the general character of the response curve will depend very much of the presupposi-

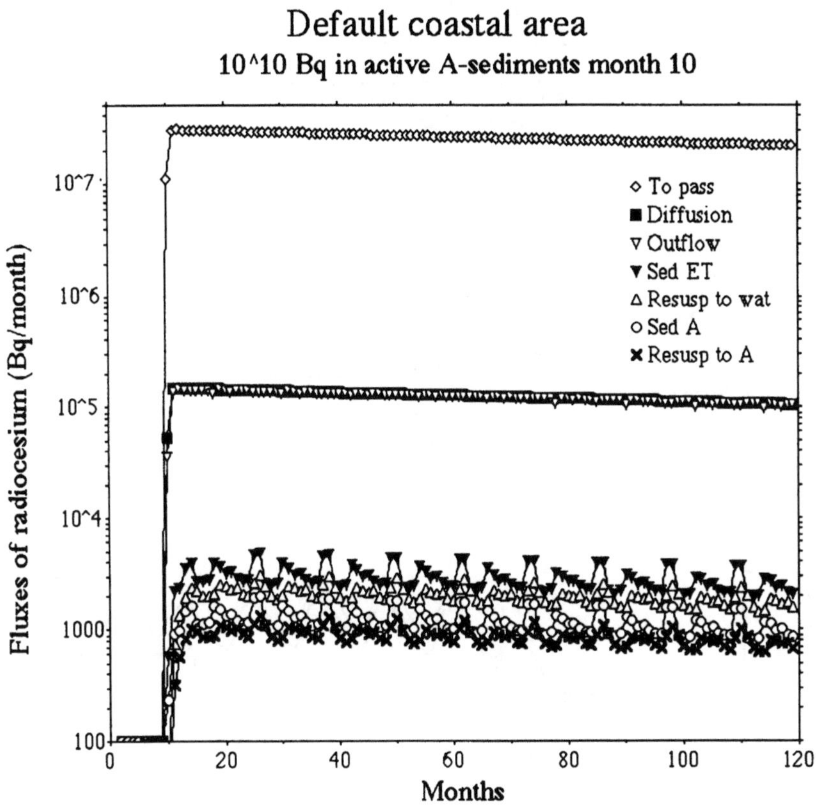

Fig. 2.30. The fluxes when 10^{10} Bq radiocesium are stored in the sediments month 10. The ranking of the fluxes is: Transport from active to passive A-sediments > diffusion > outflow > sedimentation on ET-areas > resuspension from ET-areas to water > sedimentation on A-areas > resuspension from ET to A-areas.

sitions for the Cs-concentration in the sea and the ecological halflife of ^{137}Cs in the sea. Under default conditions, however, there are no major differences in Cs-concentrations in 1-kg piscivores when the month of the fallout is changed.

2.3.2.2. Direct fallout

The direct fallout to the coastal area is simply fallout, FO (e.g., Bq/m^2) multiplied by coastal area (Area in m^2, i.e.: $F_W = FO \cdot Area$).

2.3.2.3. Inflow from the sea

The sub-model for the inflow of ^{137}Cs from the sea and/or adjacent coastal areas is given in Figure 2.33. As stressed before, it is costly and laborious to determine the theoretical (characteristic) surface and deep water retention times (Ts and Td) empirically.

The inflow is related to the measured Cs-concentration in the sea outside the coast (Cs_{sea}) and the total water exchange between the coast and the sea.

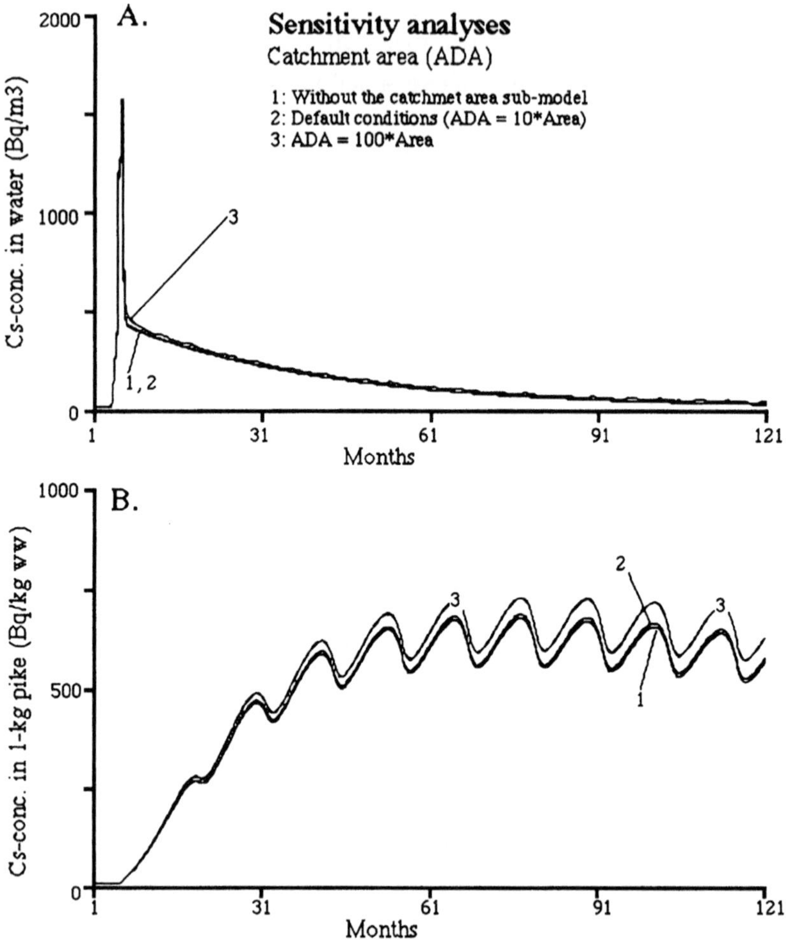

Fig. 2.31. Sensitivity tests for the catchment area sub-model in relation to radiocesium concentration in water (A) and in 1-kg piscivores (pike) (B).

The Cs-concentration outside the given coastal area is given in Bq/m^3. Cs_{sea} is a most important driving variable. Preferably, the value should be based on measurements. It is, however, evident that it would be a serious drawback if this model did not include a sub-routine to estimate Cs_{sea}. Lozán et al. (1996) have presented data from the Baltic concerning Cs-concentrations in surface water (1984-1991). There is also information about the fallout into the Baltic (see publications from HELCOM, the Helsinki commission). Fallout varied from 1 to 75 kBq over large parts of the Baltic. There is a clear, but in the details very complex, relationship between fallout and surface water concentration of ^{137}Cs. The sub-routine in this model uses a calculation constant of 0.01 (m^2/m^3) to transform the fallout (Bq) into a characteristic Cs-concentration in water (Bq/m^3) at the month of the fallout (see Figure 2.33). Since the default coastal area had

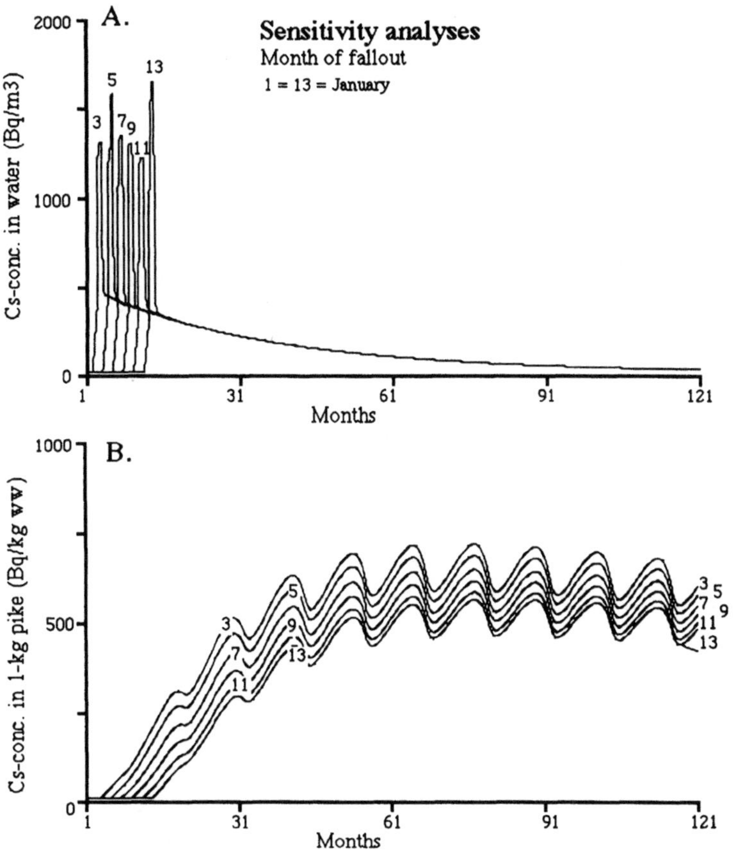

Fig. 2.32. A comparison between modelled values for total Cs-concentrations in water (A) and Cs-concentrations in 1-kg piscivores (B) (in the default coastal area) in a scenario in which the month of the fallout has been changed (fallout in January, March, May, July, September and November).

a fallout of 50 kBq/m^2, the estimated initial Cs-concentration in the sea water was 500 Bq/m^3 for the default coastal area. This value is typical for large parts of the Baltic in 1986 (see Lozán et al., 1986).

The data presented by Lozán et al. (1996) also illustrate the slow recovery process for Cs_{sea}. The Cs-concentrations even increase in large parts of the Baltic from 1986 until 1990. This model uses a simple ecological halflife approach to calculate the time-dependent changes in Cs_{sea}, as given by:

$$Cs_{sea} = (FO \cdot 0.01) \cdot e^{(-time \cdot Tsea)} \tag{2.6}$$

FO = fallout in Bq/m^2;
time = month after fallout where 1 is January in the year of the fallout;

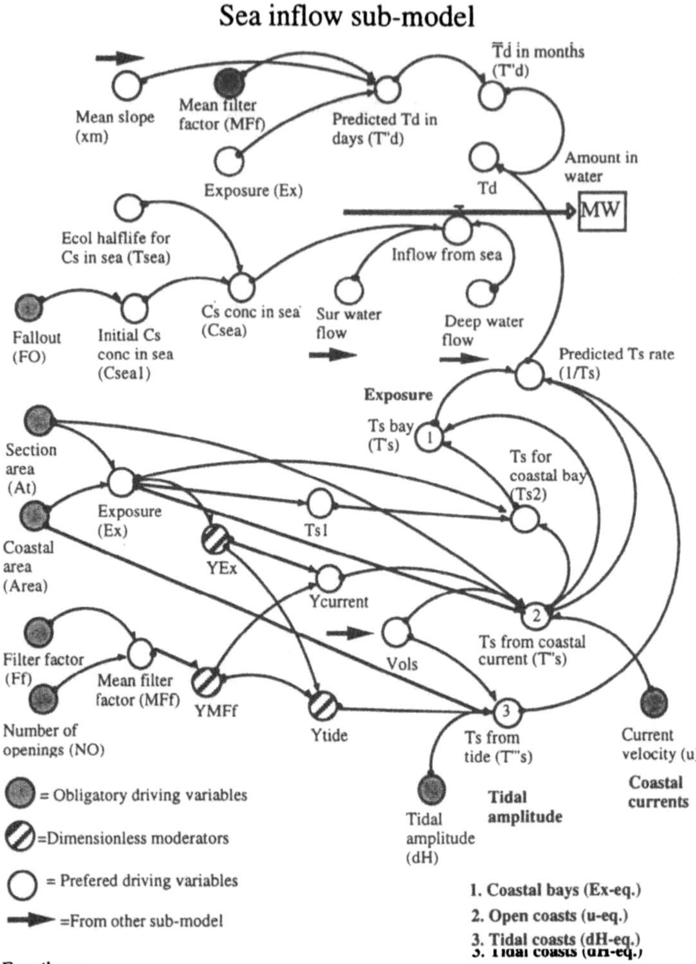

Equations:

```
YEx=(1+0.5*((Ex/10)-1))
YMFf=(1+0.5*((MFf/50)-1))
If YEx*YMFf > 1 then Ytide=1 else Ytide=YEx*YMFf
If YEx*YMFf > 1 then Ycurrent=1 else Ycurrent=YEx*YMFf
If (EXP(3.49-4.33*(Ex^0.5)))/(30) < 1/30 then Ts1=1/30 else Ts1=(EXP(3.49-4.33*(Ex^0.5)))/(30)
If Ex ≥ 1.3 then Ts2=T"s else Ts2=Ts1
If Ts2 ≥ T"s then T's=100 else T's=Ts2
If Ex < 1.3 then T"s=100 else T"s=((Vols)/(Ycurrent*u*0.01*60*60*24*30*0.5*At))
T'"s=(Vols)/(Area*dH*Ytide*0.01*30)
If EXP(7.72-2.93*(MFf)-19.16/xm-0.6*ln(Ex)) < 1 then T'd=1 else
T'd=EXP(7.72-2.93*(MFf)-19.16/xm-0.6*ln(Ex))
If1/Ts/30 > 4 then T"d=4 else T"d=T'd/30
If T"d < 3/30 then 3/30 else Td=T"d+0*10*Ts+0*100*Ts
If (1/T's)+(1/T"s)+1*(1/T'"s) > 30 then 1/Ts=30 else 1/Ts=(1/T's)+(1/T"s)+(1/T'"s)
Csea1=FO/100
Csea=Csea1*EXP(-time*Tsea)
MFf=Ff/NO
```

Fig. 2.33. An outline of the sea inflow sub-model with a sub-model for different approaches (Ex-formula, u-formula and dH-formula) for calculating the theoretical surface water retention time (Ts in months).

where

Cs_{sea} = Cs-concentration in the sea outside the given coast (Bq/m^3);

T_{sea} = the ecological halflife for radiocesium in the sea outside the given coast (1/months); the default value for T_{sea} in the Baltic is set to 3 years or 36 months. In a following section, T_{sea}-values for Atlantic and Mediterranean coasts will also be discussed.

Figure 2.34 gives sensitivity tests for different values for T_{sea}. The predicted values for Cs_{sea} (under otherwise default conditions) follow curve 1 given in Figure 2.34A. Figures 2.34B and C give the corresponding Cs-concentrations in water and 1-kg piscivores. If C_{sea} is set to 0, it illustrates a situation either when the fallout is unevenly distributed such that there is fallout onto the catchment area but not on the sea outside the coast, or, more realistically, a situation when mixing, dilution or current activity in the sea efficiently remove the fallout from the sea outside the coast. It should be stressed that the assumptions in eq. 2.6 concerning the initial Cs_{sea}-value and the value selected for T_{sea} are important for all predictions. This will be demonstrated by subsequent sensitivity and uncertainty tests. This model is based on the simplification that there are four main types of coastal areas (see Figure 2.14) where different processes regulate the surface water exchange (Ts) between the given coastal area and the sea (and/or adjacent coastal areas); see Figures 2.33 and 2.35. One of these is the freshwater inflow (Q), as already discussed. The three remaining alternatives concern the exchange processes for the surface water (Ts in months) between the given coastal area and the sea. The deep water retention is handled by yet another equation.

1. Coastal bays. In coastal areas where (a) the tidal range is smaller than about 20 cm, (b) there is no significant fresh water inflow and (c) there are no specific topographical features causing strong coastal currents (> 5 cm/s), the empirical morphometrical models for Ts and Td (see Figure 2.24) can be used to calculate the fluxes of radiocesium to and from the given coastal area. More than 90% of the coastal areas in the Baltic fall into the domain for these two models. T's (in months) is given by eq. 2.5, the Ex-formula. This model only applies to coastal areas where the exposure (Ex) is in the range from 0.002 to 1.3 and for coastal areas in the range from 0.15 to 150 km^2 and if the tidal amplitude is smaller than 20 cm/day.

For practical purposes, to avoid many time-consuming calculation steps which are necessary to obtain stable solutions using the calculation routines (Euler or Runge-Kutta) for retention rates greater than about 30 (= 30 exchanges per month), there is an "If_then_else" statement in the model which never allows the surface water retention time to less than 1 day. This is given under the symbol called "Predicted Ts rate" in Figure 2.33.

There are three alternatives in this model for estimating the deep water retention time (Td): (1) one uses a default value of 60 days, (2) Td is set to be 10 times longer than Ts (or 100·Ts for permanently stratified fjords), or (3) Td is estimated from the regression model based on three morphometric parameters (see Figure 2.24). The predicted deep water retention time (Td) is not generally allowed to be longer than 4 months, since many European coastal areas are thermally stratified during summer (and winter) and circulate during spring and autumn. For such areas, it would not be realistic

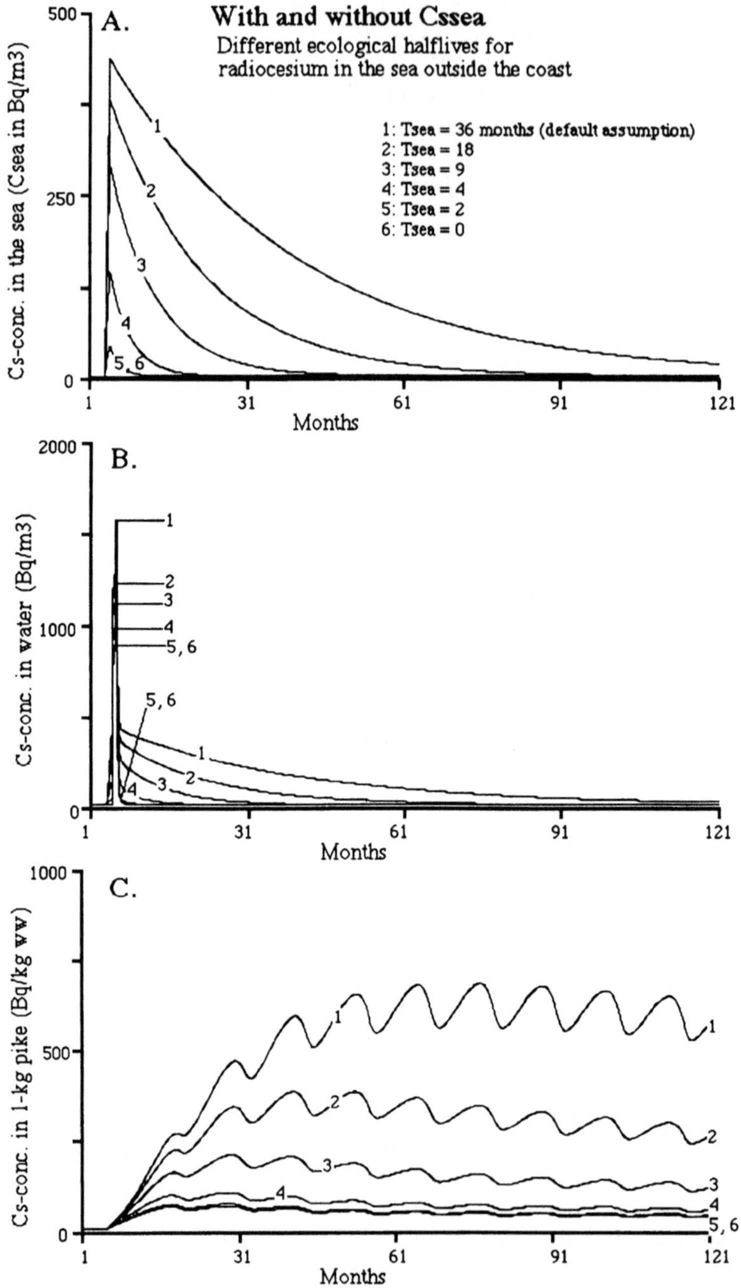

Fig. 2.34. Sensitivity tests for the ecological halflife of radiocesium in the sea outside the given coastal area in relation to (A) values for total Cs-concentrations in the sea, (B) Cs-concentrations in water in the default coastal area and (C) Cs-concentrations in 1000 g ww piscivores.

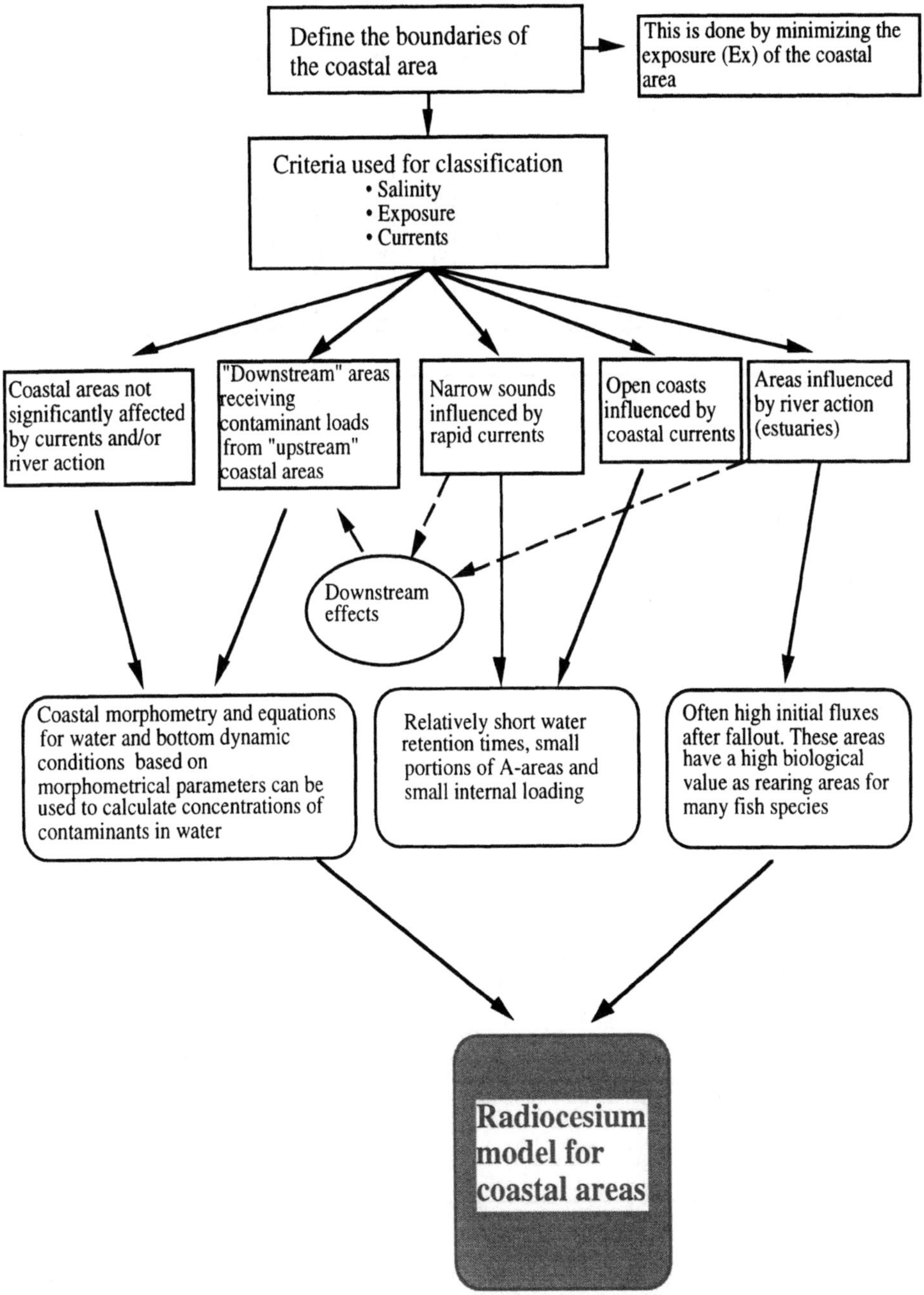

Fig. 2.35. Criteria for the equations for the surface water retention time in this model for radiocesium in coastal areas.

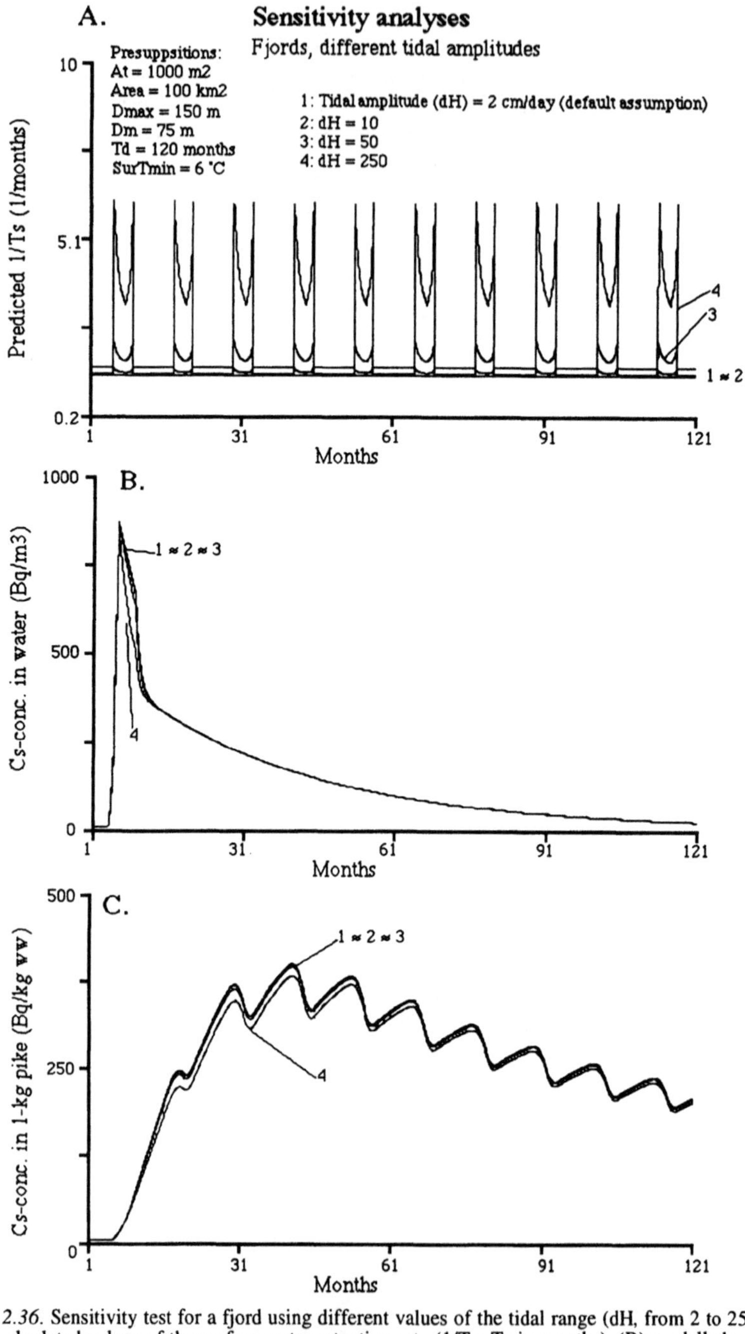

Fig. 2.36. Sensitivity test for a fjord using different values of the tidal range (dH, from 2 to 250 cm/day) for (A) calculated values of the surface water retention rate (1/Ts; Ts in months), (B) modelled values for total Cs-concentrations in water and (C) Cs-concentrations in 1000 g ww piscivores.

to allow Td to be longer than about 120 days. Thus, the conditions in stratified fjords cannot be predicted by this regression model. In the following, we will give sensitivity and uncertainty analyses to illustrate how assumptions like these influence the predictions of Cs-concentrations in water and biota.

Figure 2.36 gives one example of such a test. In this case, the model has been used to simulate the Cs-concentrations in a fjord with a narrow sill (At = 1000 m^2); the area is 100 km^2, the maximum depth (Dmax) 150 m and the mean depth (Dm) 75 m. The fjord is permanently stratified and the surface water temperature is never lower than 6 °C. The Td-value has been set to 10 years. The tidal range has been varied from 2 to 250 cm/day. Under these conditions, and when all other model variables have been kept at the default values (see Table 2.3), the model predicts a somewhat faster recovery at the higher tidal range (dH = 250 cm/days) as compared to the lowest tidal amplitude (2 cm/day), but the differences are small. The main reason for this is that even if dH is only 2 cm/day, the theoretical surface water retention time (Ts) is still less than 1 month, the water retention rate (1/Ts) is greater than 1 month, and the entire surface water mass is exchanged within one month. This explains the relatively fast initial removal of radiocesium from the surface water and the relatively fast decline in Cs_{wa}-values. The second phase of the recovery is totally dominated by the presuppositions given for the Cs-concentration in the sea outside the coast and the approach for calculating the ecological halflife for those Cs-concentrations. The processes responsible for the surface water exchange are fresh water inflow (Q) as well as coastal currents, winds, air pressure alterations, seiches, etc. accounted for in the model for Ts.

The morphometric regression for Td is given by:

$$Td = e^{(7.72-2.93\cdot(MFf)-19.16/xm-0.6\cdot\ln(Ex))}/30 \qquad (2.7)$$

where
Td = the theoretical deep water retention time (in months);
MFf = the mean filter factor (km^3);
xm = the mean slope (%);
Ex = the exposure (dim. less).

2. Open coasts. For open coasts, Ts (in months) is based on data on the characteristic monthly coastal current (u). Typical values for coastal currents in the Baltic are given in Figure 2.9. This model uses a default value of 2.5 cm/sec (= u; see Håkanson et al., 1984).

If the exposure (Ex) is smaller than 1.3, the influence of coastal currents are accounted for by eq. 2.5. Coastal currents are very important for open coastal areas. This model uses a system of simple operational rules based on two important morphometrical features, the exposure (Ex) and the mean filter factor (MFf), in setting the driving processes for the surface water exchange. The influence of coastal currents is first regulated by the exposure (Ex); if Ex > 1.3 (the limit for eq. 2.5), then the following equation based on coastal currents applies. The character of the coastal area outside the given coastal area, as given by the mean filter factor (MFf = "the energy filter"), also influences how the coastal current influences the given Ts-value. This is given by two dimensionless moderators, Y_{Ex} and Y_{MFf}, defined by (see also Figure 2.33):

$$Y_{Ex} = (1+0.5\cdot((Ex/10)-1)) \tag{2.8}$$

where
0.5 = the amplitude value; this value regulates the influence that an alteration in the actual exposure (Ex) would have relative to the norm-value (here = 10) on the target Ts-value. The norm-value of 10 means that if the exposure is larger than 10, then coastal currents (and tides) are likely to fully influence the requested Ts-value and if the actual Ex-value is lower than the norm-value, the coastal currents will not influence the Ts-value as much, i.e., the theoretical water retention time (Ts) will be longer. It is evident that this is a logical construct which needs to be tested against reliable empirical data for many coastal areas. Unfortunately, such data are difficult to access and it has been beyond the scope of this work to carry out such studies. This is a theoretical approach for this radiocesium model - nothing more, nothing less. If Ex = 1, then $Y_{Ex} = 0.55$ and coastal currents (u) influence Ts 55% less than if Ex = 10.

The dimensionless moderator for the mean filter factor, Y_{MFf}, is given by:

$$Y_{MFf} = (1+0.5\cdot((MFf/50)-1)) \tag{2.9}$$

where
0.5 = the amplitude value; the same amplitude value as in the previous moderator has been used in this case. The denser the outside archipelago, the greater the "energy filter", the smaller the influence of the coastal current. The selected norm-value of 50 means that if the MFf is larger than 50, then coastal currents (and tides) are likely to fully influence the requested Ts-value. Note that a filter factor of 100 means an open coast without any topographical obstacles like islands, reefs, etc. If MFf = 1, then Y_{MFf} = 0.495 and coastal currents (u) influence Ts 50% less than if MFf = 50.

If Ex and/or MFf are larger than the norm-values, there is an "If_than_else" statement which sets the moderators to 1. This is indicated by the symbol $Y_{current}$ (= $Y_{Ex}\cdot Y_{MFf}$) in Figure 2.33.

This means that the theoretical surface water retention time related to coastal currents (T"s in months; see Figure 2.33) is given by:

$$T''s = Vol_s/(Y_{current}\cdot u\cdot 0.01\cdot 60\cdot 60\cdot 24\cdot 30\cdot 0.5\cdot At)) \tag{2.10}$$

where
Vol_s = the volume of the surface water (m^3), which is calculated by another sub-model (which will be given in the following);
u = the characteristic coastal current; u is set to 2.5 cm/sec as a default value;
At = the section area in m^2;
0.5 = it is assumed that 50% of the section area is involved in the active water exchange.

3. Tidal coasts. For tidal coasts, Ts can be estimated from (see Håkanson et al., 1986):

Sub-models for volumes and water flows

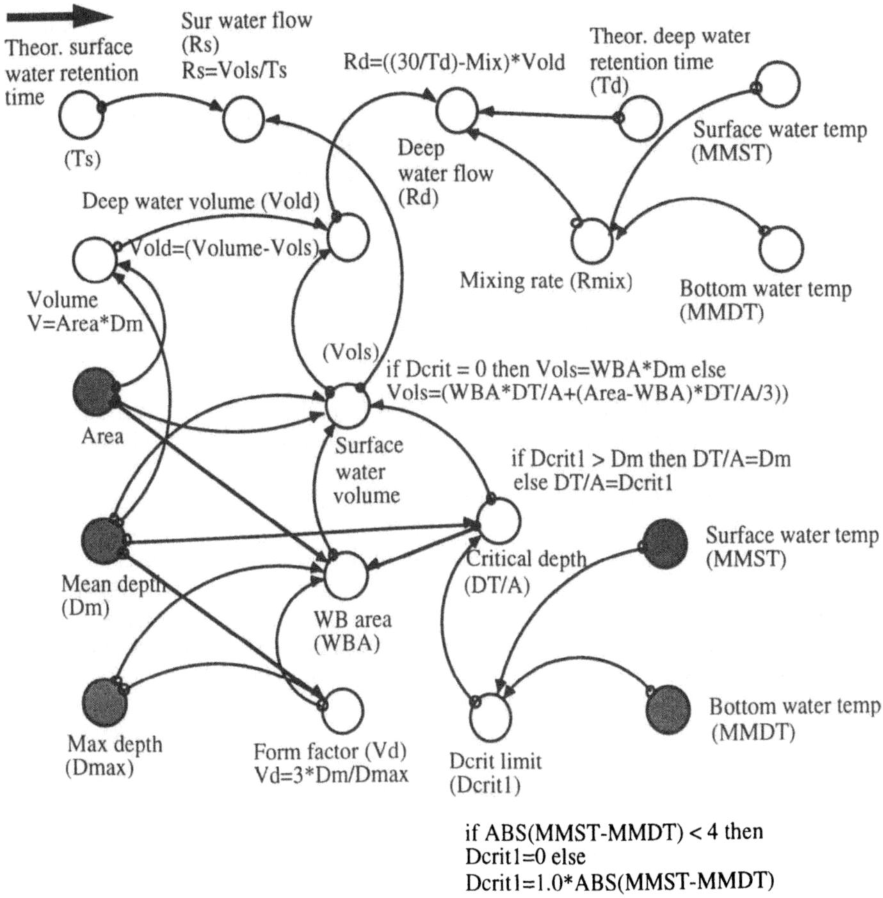

Equations:
WBA=Area*((Dmax-DT/A)/(Dmax+DT/A*EXP(3-Vd^1.5)))^(0.5/Vd)
If 1/ABS(MMST-MMDT) > 1 then Rmix=1 else Rmix=1/ABS(MMST-MMDT)

Surface water temp. (MMST in°C, assumed default values):

Jan	Feb	Mar	Apr	May	Jun	Jul	Aug	Sep	Oct	Nov	Dec
0	0	3.0	4.0	7.0	11.0	16.0	20.0	15.0	7.0	4.0	1.0

Deep water temp. (MMDT in°C, assumed default values):

Jan	Feb	Mar	Apr	May	Jun	Jul	Aug	Sep	Oct	Nov	Dec
4.0	4.0	4.0	4.0	4.5	6.0	8.0	10.0	8.0	7.0	4.0	4.0

● =Obligatory driving variables ➡ =From sub-model

Fig. 2.37. The sub-models for water volumes and water flows.

$$T'''s = Vol_s/(Area \cdot dH \cdot Y_{tide} \cdot 0.01 \cdot k \cdot 30) \quad (2.11)$$

where

Y_{tide} = $Y_{Ex} \cdot Y_{MFf}$; if $Y_{tide} > 1$ then $Y_{tide} = 1$. Y_{Ex} and Y_{MFf} are defined by equations 2.8 and 2.9;

Area = the coastal area in m^2;

k = a mixing constant; k = 1 means complete mixing. This value can be used as a default assumption if the model is run with dt = 1 month;

0.01 = a calculation constant which changes dH in cm/day to m/day;

dH = the tidal amplitude in m/month (= 30·dH, if dH is given in m/day).

4. Estuaries. There is a simple rule-of-thumb-method for estimating if it is necessary to consider the freshwater impact on the given coastal area:

- If Q is larger than 10% of the water flow from the sea, as given by:

$$R = Vol_s/(T's+T''s+T'''s) \quad (2.12)$$

where T's is given by the Ex-formula, T"s by the u-formula and T"s by the dH-formula.

- Then Q is included in the calculation and the total water flow is given by:

$$R+Q = Vol_s/(T's+T''s+T'''s) + Vol_d/(Td+Mix) + Q \quad (2.13)$$

Where the first factor is the water flow for the surface water, and the second factor is the deep water flow, which is calculated from the theoretical deep water retention time (Td), the volume of the deep water (Vol_d) and by allowing for internal mixing (see Figure 2.37), which redistributes water within the coastal area but does not contribute to the water exchange between the coastal area and the sea or adjacent coastal areas.

The volumes of the surface water and the deep water are given by the equations illustrated in Figure 2.37, which will be explained in the next section dealing with internal processes.

Figure 2.38 gives a test to illustrate how the model works and the problems with a transition from a situation in a coastal bay where the Ex-formula is used (i.e., eq. 2.5, the morphometric equation based on the exposure) to a situation in which the u-formula (i.e., coastal currents, eq. 2.10) is used. Under given default conditions (Table 2.3), the section area, and hence the Ex-value, is successively increased from Ex = 0.2 to Ex = 3.2. Eq. 2.5 should not be used if Ex > 1.3. Under default conditions, the surface water retention rate (1/Ts) is about 7, which means that the surface water is exchanged seven times in one month. A typical Ts-value is about 2-3 days, so the default conditions are intentionally less dynamic than the prevailing conditions in a typical Baltic coastal area. The corresponding Cs-concentrations in water are given in Figure 2.38B (curve 1, the highest peak and the longest tail). The next scenario, when At = 2·0.2, corresponds better to "typical" conditions in Baltic coastal areas (1/Ts ≈ 15 and Ts ≈ 2 days). This is curve 2, and one should notice the difference between the two situations during the first month after the fallout. The reason for this is that seven surface water exchanges per month is already rather large. Thus, in this model (using a dt of 1 month), it does not matter so much if the retention rate is doubled. In the next test (for Ex = 0.8), the

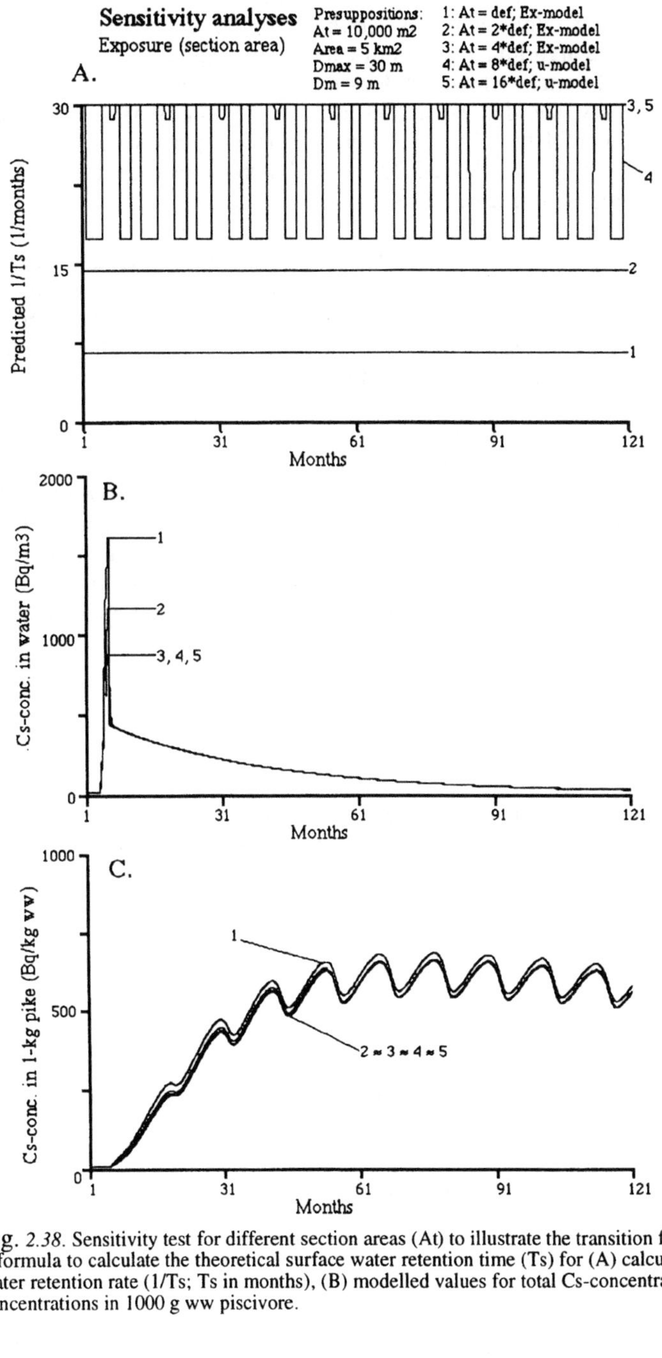

Fig. *2.38*. Sensitivity test for different section areas (At) to illustrate the transition from the Ex-formula to the u-formula to calculate the theoretical surface water retention time (Ts) for (A) calculated values of the surface water retention rate (1/Ts; Ts in months), (B) modelled values for total Cs-concentrations in water and (C) Cs-concentrations in 1000 g ww piscivore.

Ex-formula predicts T's to be less than 1 day and the practical limit for the retention rate is reached. If Ex = 1.6, the Ex-formula no longer applies and Ts is calculated using the u-formula (eq. 2.10 as T"s). This means that Ts (or rather T"s) depends on the value for the surface water volume (Vol_s), which is derived from the surface water and deep water temperatures and the algorithms for stratification (see Figure 2.37). This means that T"s shows a seasonal pattern which T's does not. T's is a characteristic constant related to the Ex-value only, and this value does not vary with the season of the year. Thus, there is a transition problem at Ex = 1.3. For all practical purposes in this model for radiocesium using dt = 1 month, Figure 2.37 shows that this transition problem does not matter very much for these target variables. However, there are situations when it is important to solve this transition problem, e.g., if dt is set to 1 day or 1 hour, and the aim is to calculate such short-term variations in Cs-concentrations in water and biota.

Figure 2.39 gives results from sensitivity tests for the given different approaches for the tidal amplitude (dH), the coastal currents (the u-formula) and the exposure formula for the theoretical surface water retention time, Ts. The higher the tidal amplitude, and the stronger the coastal currents, the quicker the water turnover time in the coastal area and the greater either the dilution or the flux of cesium from the sea to the coast. The tidal amplitude should always be accounted for if dH is larger than about 20 cm/day, and if dH is larger than about 50 cm/day it is generally the most important driving process for Ts. The influence of coastal currents is automatically taken into account by the Ex-formula, but if the current activity is known to be strong, or if the coastal area is very open (if the exposure is > 1.3), then the coastal current approach with or without the tidal approach (without the tidal approach if dH < 20 cm/day) is used. These simulations show that under default conditions, the model predictions are not so sensitive to the approach for Ts.

2.3.2.4. Internal processes

2.3.2.4.1. Water

In this model, there is a distinction between surface water and deep water in the determination of the surface water volume and the deep water volume (Figure 2.37). The fluxes of radiocesium to and from the compartment "water" are given by the following equation:

$$M_W(t) = M_W(t - dt) + (F_W + F_{AW} + F_{OAW} + F_{IAW} + F_{ETW} - F_{WO} - F_{WA} - F_{WET})\cdot dt \qquad (2.14)$$

where

M_W = amount of X in water (Bq);
F_{AW} = diffusion from A-areas (Bq/month);
F_{ETW} = advective transport (= resuspension) from ET-areas to water (Bq/month);
F_{WO} = outflow of radiocesium from water (Bq/month);
F_{WA} = sedimentation from water to A-areas (Bq/month);
F_{WET} = sedimentation from water to ET-areas (Bq/month).

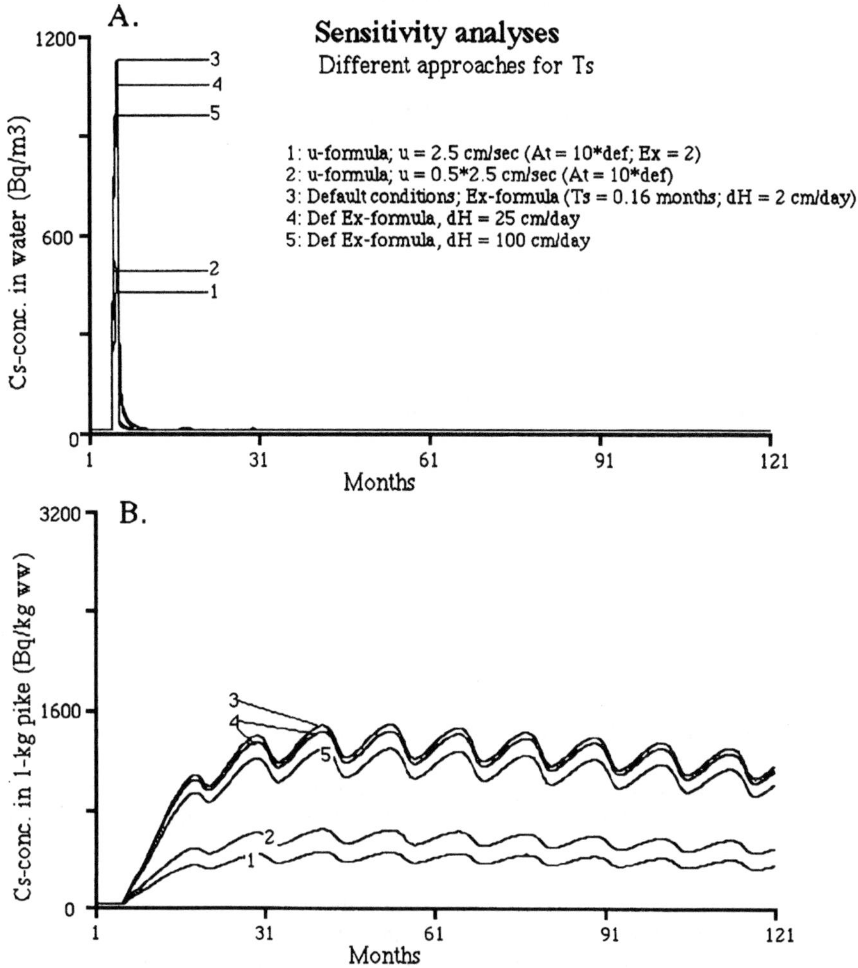

Fig. 2.39. Sensitivity test for different approaches to calculate the theoretical surface water retention time (Ts) for values for total Cs-concentrations in water (A) and Cs-concentrations in 1000 g ww piscivores (B).

The diffusive transport of radiocesium from A-areas to water, F_{AW}, is basically given by the same equations as for lakes, equations 1.17 to 1.19.

The advective transport from ET-areas to water, F_{ETW}, is given by eq. 1.23. However, for coastal areas, it is assumed (see Håkanson, 1999) that the mean age of these deposits is about 21 days. This means that the halflife, T_{ET}, is given by:

$$T_{ET} = (0.5 \cdot 21)/(0.693 \cdot 30) \tag{2.15}$$

The dissolved fraction (C_{diss}) for ^{137}Cs in coastal water is given by the same algorithm as that used for lakes (eq. 1.25).

Values of suspended particulate matter (SPM in mg/l) coastal waters should be empirically measured or predicted from a model (see Håkanson, 1999).

2.3.2.4.2. Outflow

The outflow is given by a flux called F_{WO} (from water out of the coastal area, Bq/month).

$$F_{WO} = (R_d+R_W)\cdot M_W \tag{2.16}$$

where
R_d = the physical decay;
R_W = the theoretical water retention rate (1/month);
M_W = the amount of radiocesium in the water (Bq).

R_W for coastal areas is calculated from eq. 2.13 accordingly:

$$R_W = (R+Q)/Vol \tag{2.17}$$

2.3.2.4.3. Sedimentation on A-areas

Sedimentation on coastal A-areas (F_{WA}, transport from water to A-areas) is given by:

$$F_{WA} = M_W\cdot R_{sed}\cdot(1\text{-ET}) \tag{2.18}$$

where
M_W = the amount of radiocesium in the water (Bq);
R_{sed} = the sedimentation rate (1/month);
ET = the fraction of ET-areas, which is, preferably, determined empirically (see Håkanson and Jansson, 1993, for methods) or predicted from a sub-model (see Figure 2.24).

The sedimentation rate for particulate radiocesium, R_{sed}, is given by eq. 1.33. Figure 2.40 illustrates the sub-model for sedimentation of particulate radiocesium, sedimentation of matter (gross sedimentation) and the calculation of the age of the active A-sediments.

As noted for the lake model, and as discussed in connection with equations 1.34 and 1.35, SPM will be allowed to influence the settling velocity for particulate radiocesium. This is achieved by means of a dimensionless moderator, Y_{SPM}. The amplitude value for this moderator for the coastal model is set in such a manner that a change in the concentration of SPM by a factor of 10, e.g., from 3 (which is a typical value for the Baltic; see Pustelnikov, 1977) to 30 mg/l (which would be typical of highly productive coastal areas), will cause a change in the settling velocity by a factor of 3. The normal value for the moderator is set to 3 mg/l. The default settling velocity for particulate radiocesium is set to 12 m/yr (from Håkanson, 1999). Thus, v_{Cs} is given by:

$$v_{Cs} = (12/12)\cdot Y_{SPM}\cdot Y_{Sal}\cdot((1\text{-}D_{res}) + 2\cdot D_{res}) \tag{2.19}$$

Sedimentation and sediment age sub-model

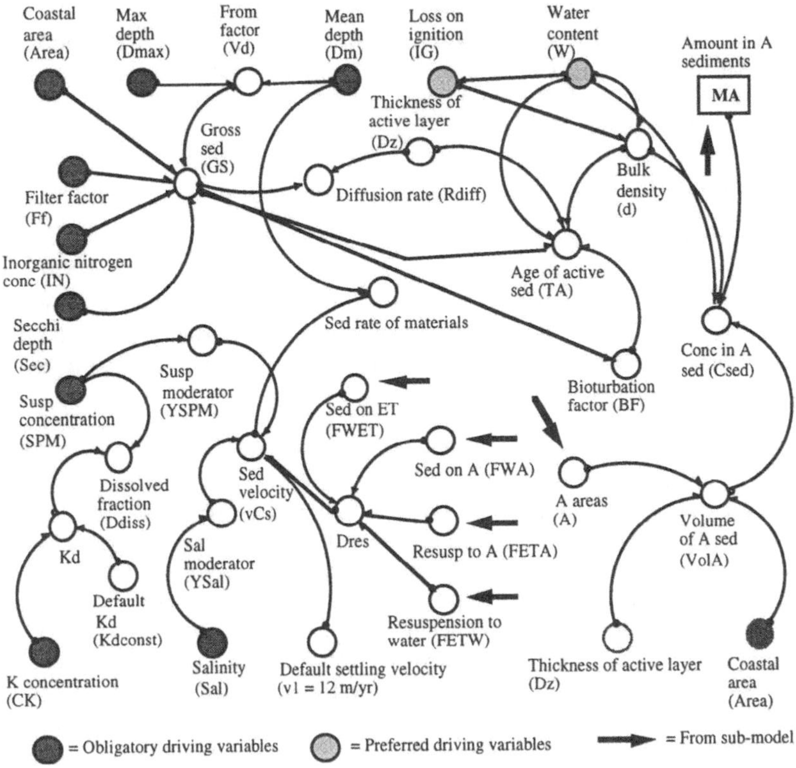

Equations:
Kd=Kdconst/CK
Ddiss=1/(1+(Kd*SPM)/1000000)
YSPM=1+0.22*(SPM/3-1)
YSal=(1+0.1*(Sal/1-1))
Dres=(FETW+FETA)/(FWA+FWET)
If Dres > 0.99 then vCs=v1*2*YSPM*YSal else vCs=YSPM*YSal*(v1*(1-Dres)+2*v1*Dres)
If ((Vd/3)*10^(0.323-0.611*log(Area*10^(-6))+1.009*log(IN)-0.587*log(Sec)+0.0238*Ff)) < 1 then GS=1 else
GS=((Vd/3)*10^(0.323-0.611*log(Area*10^(-6))+1.009*log(IN)-0.587*log(Sec)+0.0238*Ff))
Rdiff=(0.002/12)*(GS/10)*(0.02/Dz)
d=100*2.6/(100+(W+IG')*(2.6-1))
TA=BF*Dz*(1-W/100)*d*10^6/(GS*30)
If GS > 20 then BF=1 else BF=(4-0.02*(GS/1-1))
VolA=Area*(A/100)*Dz
Csed=MA/((10^3)*VolA*d*(1-W/100))

Fig. 2.40. The sub-model for sedimentation of particulate radiocesium, sedimentation of matter (gross sedimentation) and the calculation of the age of the active A-sediments.

where the dimensionless moderator expressing the influence of SPM on the fall velocity, Y_{SPM}, is given by:

$$Y_{SPM} = (1+0.22\cdot(SPM/3-1)) \quad (2.20)$$

The salinity of the coastal water will also influence the settling velocity of particulate radiocesium: the higher the salinity, the greater the aggregation of suspended particles, the bigger the flocs and the faster the settling velocity (Kranck, 1973, 1979; Lick et al., 1992). This is expressed by a dimensionless moderator (Y_{Sal}) operating on the default settling velocity. The salinity is given in ‰. The norm-value of the moderator is 1‰ and the amplitude value is set to 0.1 (see Håkanson, 1999). This means that if the salinity changes from 5 to 10‰, the moderator (Y_{Sal}) changes from 1.4 to 1.9 and the sedimentation rate increases by a factor of 1.9/1.4 = 1.36.

This means that Y_{Sal} is given by:

$$Y_{Sal} = (1+0.1\cdot(Sal/1-1) \qquad (2.21)$$

The settling velocity of particulate radiocesium also depends on the amount of resuspended matter. As for lakes, it is assumed that resuspended particles will settle out two times faster than the primary materials. The fraction of resuspended matter is given by means of the distribution coefficient (D_{res}):

$$D_{res} = (F_{ETA}+F_{ETW})/(F_{WA}+F_{WET}) \qquad (2.22)$$

It would be unrealistic to assume that the same default SPM-value of 3 mg/l could be used for ALL coastal areas. The value selected as the norm-value (3) is typical for the Baltic. The amplitude value is 0.22. This means that if SPM is 6, Y_{SPM} = 1.22 and the settling of particulate radiocesium is 1.22 times faster than under default conditions. There is also a logical negative relationship between Secchi depth and SPM; see Figure 2.41. The more suspended particulate matter in the water, the lower the Secchi depth in the given coastal area. The r^2-value for the regression in Figure 2.41 is 0.81 (n = 26). The regression is based on mean monthly data from coastal areas in the Åland archipelago from an ongoing study (Lennart Nordvarg, pers. comm.).

The amount of suspended particulate matter (SPM) is important, not so much for the total Cs-concentrations in water, but certainly for the Cs-concentrations in biota. Figure 2.42 gives sensitivity analyses for the concentration of suspended particulate matter. The default value for the concentration of suspended particulate matter has been multiplied by 0.1 and 10 to illustrate how this would influence the radiocesium concentration in water and in 1000 g ww piscivores. The more suspended matter, the lower the dissolved fraction of radiocesium, the greater the sedimentation of particulate radiocesium, the lower the concentrations in water and the lower the concentrations in planktivores, e.g., small perch, and the higher the Cs-concentrations in benthivores and, as illustrated in Figure 2.42B, in piscivores. This sensitivity analysis demonstrates that it is important to have reliable empirical data for the amount of suspended matter in the coastal area, since this influences fundamental processes regulating the spread and uptake of radiocesium, e.g., the dissolved fraction, which regulates the relationship between pelagic and benthic pathways for radiocesium, and the settling velocity, and hence, how much of the radiocesium will be retained within the given coastal area. This is one of the most important obligatory driving variables, especially in predicting Cs-concentrations in benthivores, piscivores and omnivores. Uncertainties in this value will imply uncertainties in all predictions. Unfortunately, at present there seem to be no

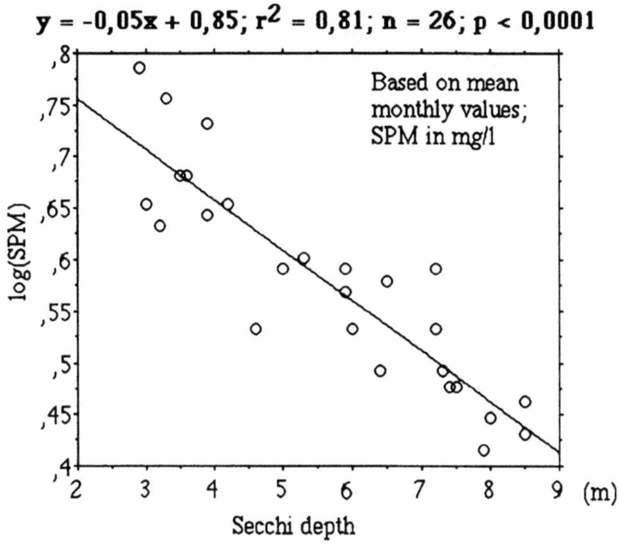

Fig. 2.41. Regression between mean monthly values fort Secchi depth and suspended particulate matter concentration (SPM) from coastal areas in the Åland archipelago (data from Lennart Nordvarg, pers. comm.).

simple models for predicting the amount of suspended matter (see Håkanson, 1999), but realistic order of magnitude values may be obtained from Table 2.5 for different coast types. If data on Secchi depth are available, the regression between Secchi depth and SPM given in Figure 2.41 may be used to estimate a SPM-value.

Table 2.5. Characteristic features in coastal areas of different trophic levels in the Baltic. Note that there is a great overlap between the different categories, such that in oligotrophic areas the concentrations of total-N may vary within a year from very low to high values. All variables are expressed as summer mean values (modified from Wallin et al., 1992).

Trophic level	**Secchi** (m)	**Chl-a** (mg/m^3)	**Total-N** (mg/m^3)	**Inorg-N** (mg/m^3)	**SedS** ($g/m^2.d$)	**O_2B** (mg/l)	**O_2Sat** (%)	**SPM** (mg/l)
Oligot.	6	<2	<280	<15	<3	>9	>90	<3
Mesot.	3-6	1-4	260-350	10-35	2-12	5-10	50-100	2-5
Eut.	1.5-3	3-5	300-450	30-45	10-18	4-6	40-60	4-10
Hypert.	<1.5	>5	>400	>40	>15	<4	<40	>6

SedS = sedimentation in near-surface sediment traps; O_2B = oxygen concentration in the bottom (= deep) water; O_2Sat = oxygen saturation in the deep water; SPM = suspended particulate matter.

These results concerning the role of SPM are important and can be compared to similar results for the fall velocity (v_{Cs}). Figure 2.43 gives sensitivity tests for the settling velocity of particulate radiocesium. The default value for the settling velocity is set to 12 m/yr. This value is based on well tested results for lakes (see section 1). In this simulation, this default value has been multiplied by 0.1 and 10. The figure illustrates how this would influence the radiocesium concentration in water and in 1000 g ww piscivores. This sensitivity analysis demonstrates that it is important to have relevant data for the settling velocity. This is a most important for empirical model calibrations.

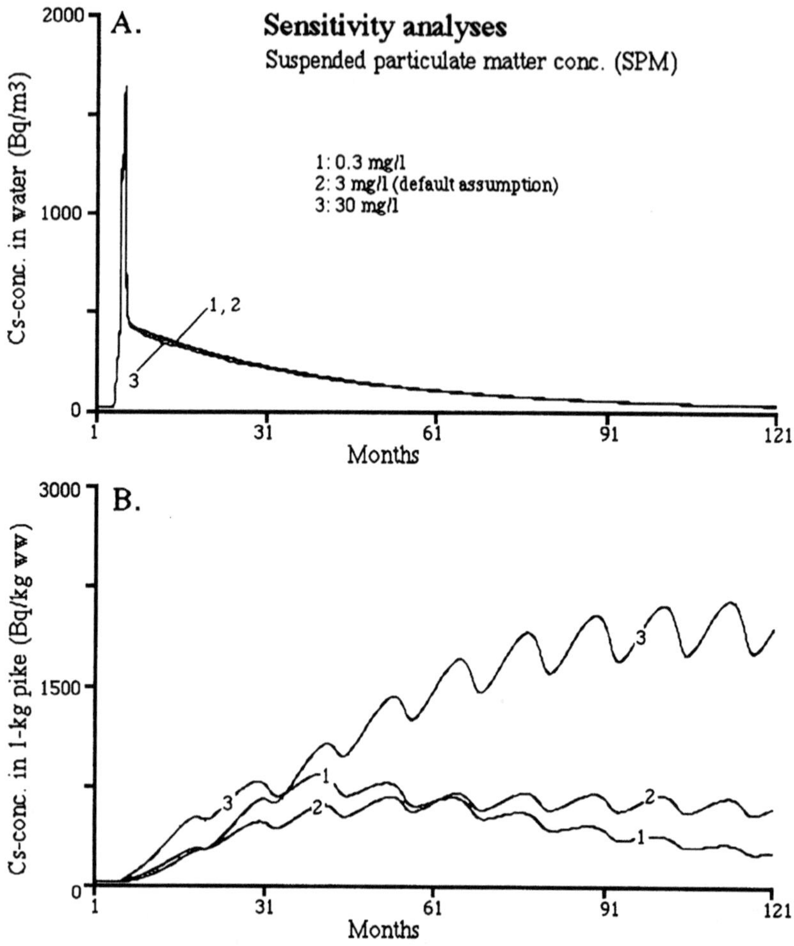

Fig. 2.42. Sensitivity test for suspended particulate matter (SPM) for values for total Cs-concentrations in water (A) and Cs-concentrations in 1000 g ww piscivores (B).

The settling velocity is influenced (1) by the salinity - the higher the salinity, the greater the settling velocity, (2) by the amount of suspended matter - the more suspended matter, the faster the settling velocity, and (3) by the amount of resuspended matter - the resuspended matter has a default settling rate which is twice as fast as that of primary materials. The sedimentation rate is then given as the ratio between the settling velocity and the mean depth. To calculate the particulate fraction one also needs data on the potassium concentration. The high settling velocity of 120 m/yr means that large quantities of radiocesium settle out in the coastal area and cause high exposure in benthivores and piscivores.

Figure 2.44 gives sensitivity analyses for salinity. The default value for the salinity is 6‰ (for the Baltic). In this simulation, the salinity was varied from 0 to 24‰. The figure illustrates how this would influence the Cs-concentration in water, in 10

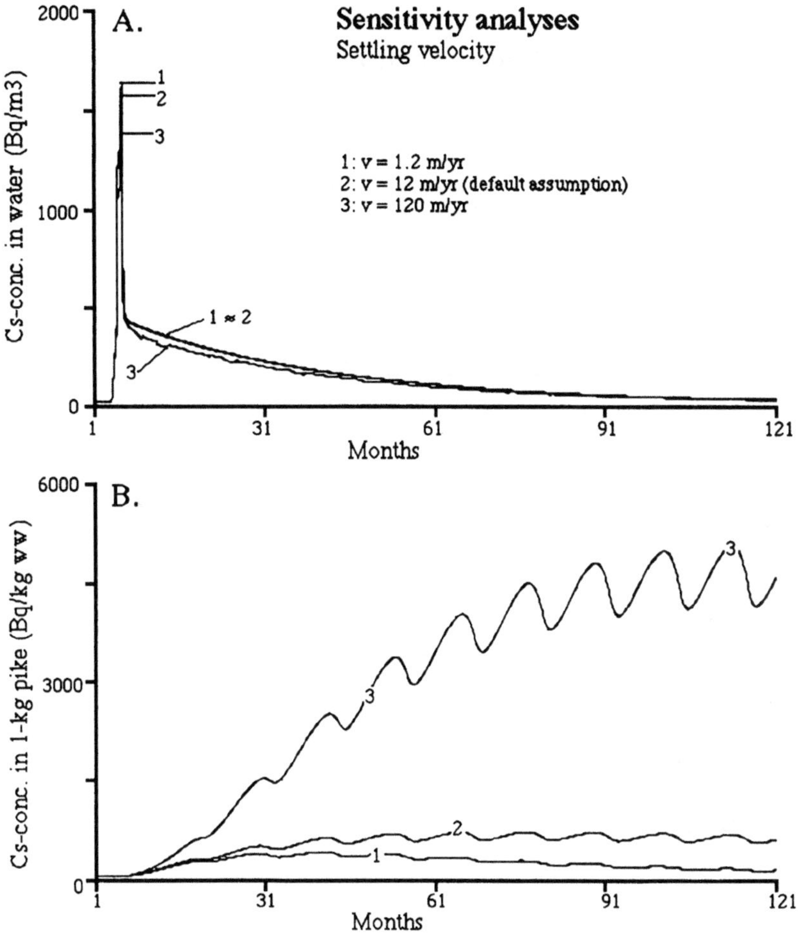

Fig. 2.43. Sensitivity test for the settling velocity for particulate radiocesium for values for total Cs-concentrations in water (A) and Cs-concentrations in 1000 g ww piscivores (B).

G ww planktivores and in 1-kg piscivores. The salinity influences both the biouptake (the higher the salinity the lower the biouptake), and the settling velocity of the carrier particles for the particulate ^{137}Cs (the higher the salinity the faster the settling velocity). This means that an increase in salinity means lower expected Cs-concentrations in water (if all else is constant), potentially lower Cs-concentrations in planktivores, but higher Cs-values in piscivores (and benthivores). In the default coastal area there are, however, no major differences in the Cs-concentrations in water, but the Cs-values in piscivores actually increase when the salinity increases from 12 to 24‰. In fresh water and brackish water, the influence of salinity on the settling process is not big enough to compensate for the influence of salinity on the biouptake so the predicted Cs-values in fish decrease when the salinity increases in the range from 0 to about 10‰. These sensitivity analyses demonstrate that it is very important to have reliable empirical data on the salinity in the given coastal area, since this value influences two important

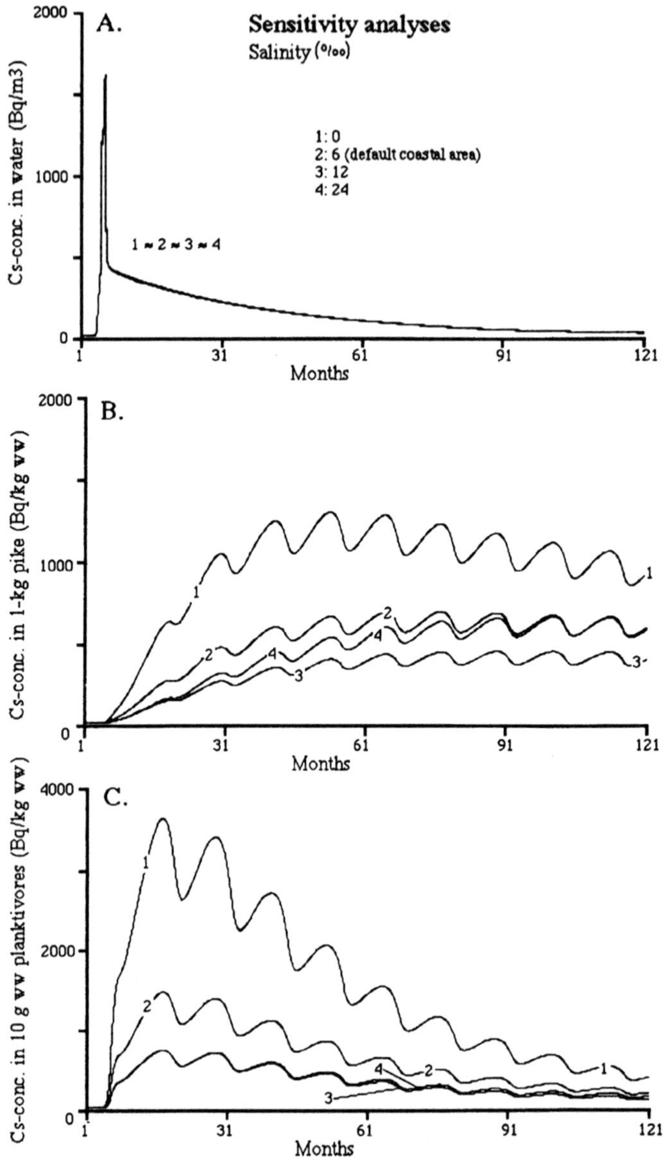

Fig. 2.44. Sensitivity test for the salinity in relation to values for total Cs-concentrations in water (A), Cs-concentrations in 1000 g ww piscivores (B) and Cs-concentrations in 10 g ww planktivores (C).

processes regulating the spread and sedimentation of particulate radiocesium and the biouptake. It is important to carefully test the salinity moderator regulating the effect of the salinity on the biouptake. Also, the algorithm describing how the salinity is supposed to influence the settling velocity is uncertain.

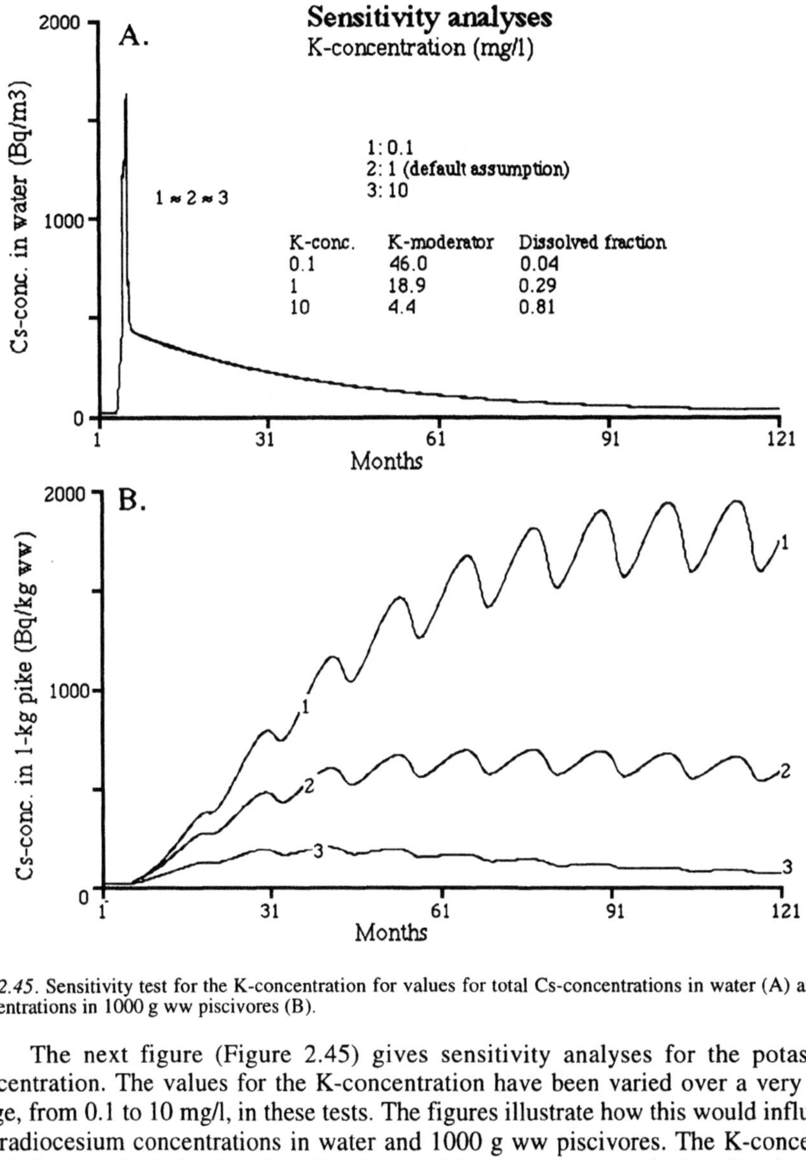

Fig. 2.45. Sensitivity test for the K-concentration for values for total Cs-concentrations in water (A) and Cs-concentrations in 1000 g ww piscivores (B).

The next figure (Figure 2.45) gives sensitivity analyses for the potassium concentration. The values for the K-concentration have been varied over a very wide range, from 0.1 to 10 mg/l, in these tests. The figures illustrate how this would influence the radiocesium concentrations in water and 1000 g ww piscivores. The K-concentration influences both the dissolved fraction (and hence also the particulate fraction) and the biouptake. It is interesting to note that these two modes of operation may counteract one another. This is illustrated by the data in the Figure 2.45 for the K-moderator: a higher value means a higher biouptake, and, for the dissolved fraction, a higher value means that more ^{137}Cs can be transported out of the coastal area. These simulations indicate that it is important to have access to reliable data on the K-concentration, because this will influence two fundamental fluxes, i.e., sedimentation/retention and biouptake. If the K-concentration is very high, this means that more radiocesium will be retained in the coastal system and that the total concentrations in water will be relatively high; however, the concentration in biota will be comparatively low, and vice versa if the K-concentration is low.

2.3.2.4.4. Resuspension from ET-areas into water

The advective transport of radiocesium from ET-areas, either back to water or to A-areas is given by the same basic equation as that used for lakes (eq. 1.39).

2.3.2.4.5. Active A-sediments

Eq. 1.39 gives all fluxes to and from the compartment active A-sediments. The age of the active A-sediments is needed to calculate the retention rate of the substance in this compartment (see Figure 2.40), and, hence, the flux out of the compartment to passive sediments (the retention rate is set to 1/age). The age is generally calculated as the ratio between the depth of the active sediments (in m) and the deposition of materials on A-areas (in m/month). This model assumes that the default value for the depth of the active layer can be set at 5 cm (Håkanson, 1999). Gross sedimentation is calculated using an empirical model, which will be described in this section.

This section presents a new empirical model for sedimentation in coastal areas. It is partly based on the same empirical data-set used by Wallin et al. (1992). One strict demand on this model is that it must only be driven by variables already included in the overall model for radiocesium for other purposes. This means that Secchi depth, which is included only in this sub-model for gross sedimentation, is not an obligatory driving variable. Values of Secchi depth can be predicted from the regression in Figure 2.41, i.e., from measured data on SPM.

The data used to derive the empirical model for gross sedimentation incorporated in this overall model for radiocesium are given in Tables 2.7 and 2.8. The model for gross sedimentation is presented in Table 2.6. The method used to derive this empirical model, including the stepwise multiple regression analysis, is given in Håkanson and Peters (1995). This model is based for data on gross sedimentation from sediment traps placed in the deepest parts of the coastal areas (SedB). Thus, the equation provides maximum values for gross sedimentation. It is well known (see Håkanson and Jansson, 1983) that sedimentation increases from about 0 at the limit between areas of transport and areas of accumulation to maximum values at the deepest part of the basin, a phenomenon often referred to as "sediment focusing". In this approach there is a correction factor applied to the model-predicted values based on this knowledge. The predicted values for gross sedimentation are assumed to be correct for U-shaped basins, with a form factor of 3, and too low for V-shaped basins with a form factor smaller than 3. Thus, the ratio Vd/3 is used as a correction factor.

This means that the model for gross sedimentation is:

$$GS = (Vd/3) \cdot SedB \qquad (2.23)$$

where SedB = the gross sedimentation in near-bottom sediment traps, as determined from the empirical model:

Table 2.6. The "ladder" for the empirical model for predicting gross sedimentation. These results are based on data for 23 Baltic coastal areas presented by Wallin and Håkanson (1992) and methods presented by Håkanson and Peters (1995). Note that the model may produce very uncertain predictions if it is used for other coastal areas, e.g., areas influenced by heavy tides. y = log(Gross sed); Gross sedimentation in g dw/m^2·day. F ≥ 2. IN = inorganic nitrogen in mg/m^3; Ff. = filter factor in km^3; Area in km^2; Secchi depth in m.

Step	Model variable	r^2-value	Model
1	log(IN)	0.50	$y=-0.78+1.47\cdot x1$
2	log(Area)	0.61	$y=-0.48+1.40\cdot x1-0.37\cdot x2$
3	Ff	0.67	$y=-0.66+1.53\cdot x1-0.59\cdot x2+0.017\cdot x3$
4	log(Secchi)	0.74	$y=0.323+1.009\cdot x1-0.611\cdot x2+0.0238\cdot x3-0.587\cdot x4$

$$SedB = 10^{(0.323-0.611\cdot\log(Area)+1.01\cdot\log(IN)-0.587\cdot\log(Sec)+0.0238\cdot Ff))} \tag{2.24}$$

Note that Area is given in km^2 in this model. This empirical model gave an r^2-value of 0.74 for the following ranges of the model variables:

	Area (km^2)	**IN** (µg/l)	**Sec** (m)	**Ff** (km^3)
Min.	1.6	14	1	0.06
Max.	14.1	52	6.9	30.7

The values for IN (inorganic nitrogen) and Sec (Secchi depth) should be the mean, characteristic values for the production period.

So, gross sedimentation increases with the following factors.

1. Inorganic nitrogen (IN) concentration, which is related to primary production.
2. Area, which is directly related to effective fetch and the internal loading (resuspension). The larger the coastal area, the greater the fetch and the larger the resuspension, and the larger the gross sedimentation.
3. The filter factor, i.e., the "wind/wave energy filter"; the larger the Ff-value, the more open the coast.
4. Secchi depth, which evidently (see Figure 2.41) is related to the amount of suspended particulate matter in the coastal water.

If gross sedimentation (GS) is larger than 20 g dw/m^2·day, it is assumed that the oxygen consumption is so high that the oxygen concentration is smaller than 2 mg/l and bioturbation is halted because meio- and macrobenthos die.

The bioturbation factor (BF) also in this model is given by eq. 1.43.

These factors, i.e., gross sedimentation, bioturbation and compaction, will influence the age of the active A-sediments (see Figure 2.40). This means that the flow of radiocesium from active to passive A-sediments, F_{APS}, is given by eq. 1.44.

The concentration of radiocesium in active A-sediment can be calculated in a standard fashion. This requires data on the volume of the active A-sediments, the sediment water content and the sediment bulk density, as given in Figure 2.40.

Table 2.7. Data for the 23 coastal areas. * Measured and modelled Td-values from Persson and Håkanson (1996) and Håkanson (1999).

Area	Code	Area (km^2)	Dmax (m)	Dm (m)	At (km^2)	xm (%)	Ex	Ff (km^3)	MFf (km^3)	Ts (d)	Td* (d)
Lilla Rimmö	SE1	2.59	17.6	8.3	0.0172	4.09	0.67	5.32	0.89	1.0	2.6
Eknön	SE2	14.04	19.5	8.5	0.0168	2.63	0.13	4.65	0.78	7.3	2.4
Lagnöstr.	SE3	5.41	20.1	3.8	0.0032	2.86	0.07	0.58	0.29	3.0	2.1
Gräsmarö	SE4	14.15	46.9	13.8	0.0825	6.01	0.63	30.71	2.36	1.1	0.0
Ålön	SE5	6.54	35.2	8.0	0.0162	7.25	0.30	5.56	0.70	2.0	11.9
Matvik	SS1	3.12	14.3	5.2	0.0067	3.46	0.22	3.68	0.46	2.0	11.1
Boköfjärd	SS2	7.05	21.6	7.1	0.0141	3.94	0.24	17.54	1.60	4.1	0.1
Tärnö	SS3	1.54	11.1	5.1	0.0062	3.18	0.42	3.54	0.59	2.1	4.2
Guövik	SS4	2.86	22.8	5.2	0.0074	3.86	0.28	5.71	1.90	3.5	0.0
Järnavik	SS5	3.49	18.6	5.7	0.0081	4.31	0.26	5.63	1.41	3.8	0.3
Guav+Järnav.	SS6	8.71	22.8	5.0	0.0108	3.81	0.15	9.84	1.97	6.4	0.0
Spjutsö	SS7	3.49	15.6	5.8	0.0188	3.74	0.54	8.79	1.46	1.4	0.1
Ronneby	SS8	11.94	17.6	4.3	0.0176	2.21	0.16	19.46	6.49	4.0	0.0
Käldöfjärd	F1	3.14	16.7	7.6	0.0040	6.34	0.15	0.61	0.10	6.5	120
Haverö N+S	F2	2.54	22.5	8.6	0.0172	5.56	0.70	7.75	0.97	0.9	5.0
Häverö N	F3	1.47	22.5	7.8	0.0176	5.47	1.27	7.26	1.21	0.2	0.3
Haverö S	F4	1.05	19.7	9.7	0.0069	5.7	0.64	2.06	0.52	1.1	27.3
Hämmärös.	F5	2.21	19.3	7.9	0.0114	6.19	0.52	4.07	1.02	1.5	5.8
Laitsalmi	F6	4.28	18.5	7.6	0.0080	5.61	0.19	1.79	0.60	5.1	120
Särkens. N+S	F7	3.69	38.0	11.3	0.0145	6.82	0.40	4.61	2.30	2.2	0.0
Särkens. N	F8	2.06	26.8	10.0	0.0090	5.78	0.45	2.00	1.00	1.9	4.0
Särkens. S	F9	1.58	38.0	13.1	0.0090	8.17	0.56	3.50	1.75	1.3	0.1
Kaukolanlahti	F10	1.38	13.3	4.8	0.0006	4.15	0.05	0.06	0.01	14	120

Figure 2.46 gives sensitivity analyses for the influence of gross sedimentation on the radiocesium concentrations in water and 1-kg piscivores. Gross sedimentation influences the age of the active A-sediments, and hence the amount in A-sediments and the fluxes from A-sediments, i.e., diffusion and transport to passive A-sediments. Since these fluxes, as discussed, are not normally very important, as compared to the primary fluxes, one would expect the model to be rather insensitive to the choice of value for gross sedimentation, and this is also what these tests show. Note that the sediments go

Table 2.8. More data for the 23 coastal areas.

Area	ET (%)	TP (μg/l)	TN (μg/l)	IN (μg/l)	SedB (g dw/m^2·d)	Secchi (m)	Chl (mg/m^3)	O_2conc (mg/l)	O_2Sat (%)
Lilla Rimmö	0.73	21	355	26	20.205	2.08	2.26	8.26	77.2
Eknön	0.19	25	385	25	5.313	1.09	3.53	6.17	60.1
Lagnöströmmar	0.92	27	417	35	22.723	1.07	4.59	7.95	79.5
Gräsmarö	0.60	16	354	24	18.328	4.03	2.65	7.92	66.6
Ålön	0.94	14	334	25	9.511	4.08	2.05	9.01	85.3
Matvik	1.00	19	273	17	12.732	5.06	1.39	10	97.8
Boköfjärd	0.96	18	258	17	6.735	5.07	1.42	10	97.9
Tärnö	1.00	17	282	14	7.591	5.02	1.56	9.75	96.7
Guövik	0.65	21	296	15	6.382	5.03	1.97	9.09	98.1
Järnavik	1.00	23	286	16	8.081	5.02	1.26	9.08	95.7
Guav. + Järnavik	0.90	22	299	17	7.302	5.05	1.57	9.08	96.8
Spjutsö	0.99	27	256	16	8.671	6.09	0.09	9.46	91.4
Ronneby	0.78	25	304	20	20.165	4.05	2.09	9.05	92.3
Käldöfjärd	0.68	22	361	26	20.012	2.05	2.68	2.02	17.06
Haverö N+S	0.71	27	325	25	38.697	3.04	2.14	4.53	38.4
Häverö N	0.73	23	346	25	63.177	4.01	2.33	8.01	68.4
Haverö S	0.66	31	305	24	23.389	3.04	1.97	1.05	8.05
Hämmärösalmi	0.98	23	337	46	13.372	3.01	2.02	4.87	46.9
Laitsalmi	0.59	29	349	52	56.435	1.06	2.68	2.82	26.2
Särkensalmi N+S	0.96	23	401	45	67.839	1.04	4.24	6.78	63.5
Särkensalmi N	0.94	21	392	42	82.534	1.01	4.49	5.64	48.2
Särkensalmi S	0.97	26	413	48	49.089	1.06	3.91	7.35	71.1
Kaukolanlahti	0.64	24	385	18	26.341	1	9.06	1.71	16.01

anoxic if gross sedimentation exceeds 20 g dw/m^2·day. This means that the age of the sediments will be influenced not by bioturbation but by sedimentation and compaction. This, in turn, means a much faster recovery for piscivores, as shown by curve 3 in Figure 2.46B.

2.3.2.4.6. Areas of erosion and transport

The differential equation used for lakes (eq. 1.45) is also used for coastal areas.

Hence, all biotic fluxes are explained (see Figure 2.47 for a compilation), and we can turn to the biological part of the model.

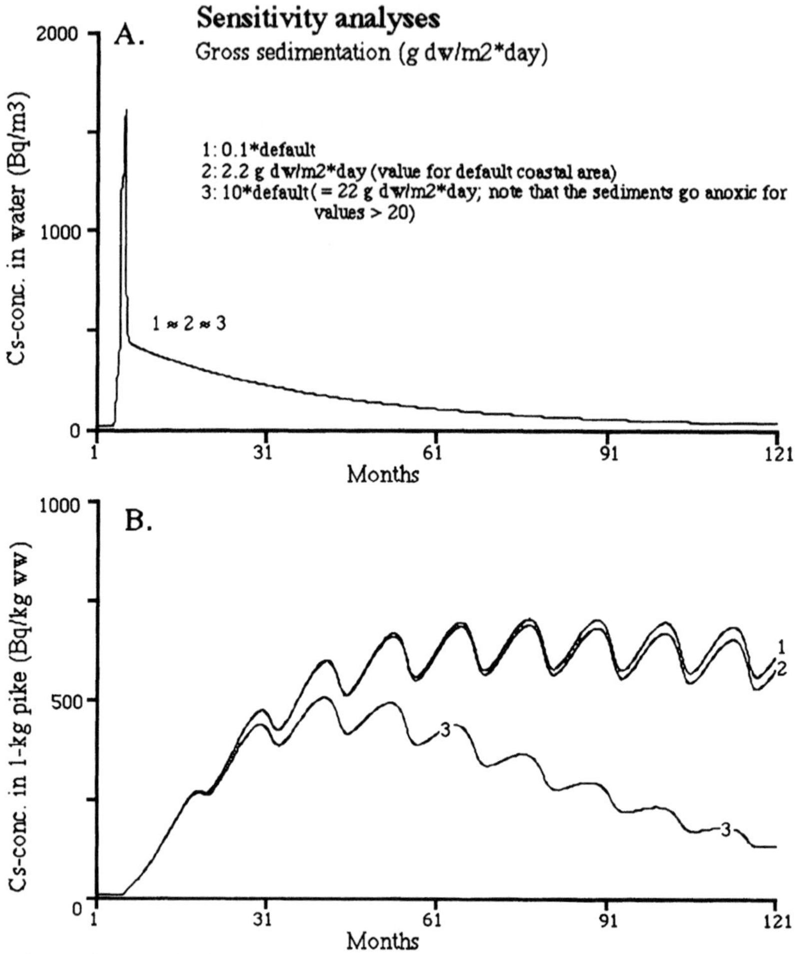

Fig. 2.46. Sensitivity test for the gross sedimentation for values for total Cs-concentrations in water (A) and Cs-concentrations in 1000 g ww piscivores (B).

2.3.2.5. Biotic processes

2.3.2.5.1. General set-up

The biotic processes are treated separately from the biotic fluxes and the two pathways are linked by means of a bioconcentration factor (Figure 1.1).

2.3.2.5.2. The fish model

The fish model for coastal areas is basically the same as the lake model, but there are some important alterations (compare Figure 1.27 with Figure 2.48).

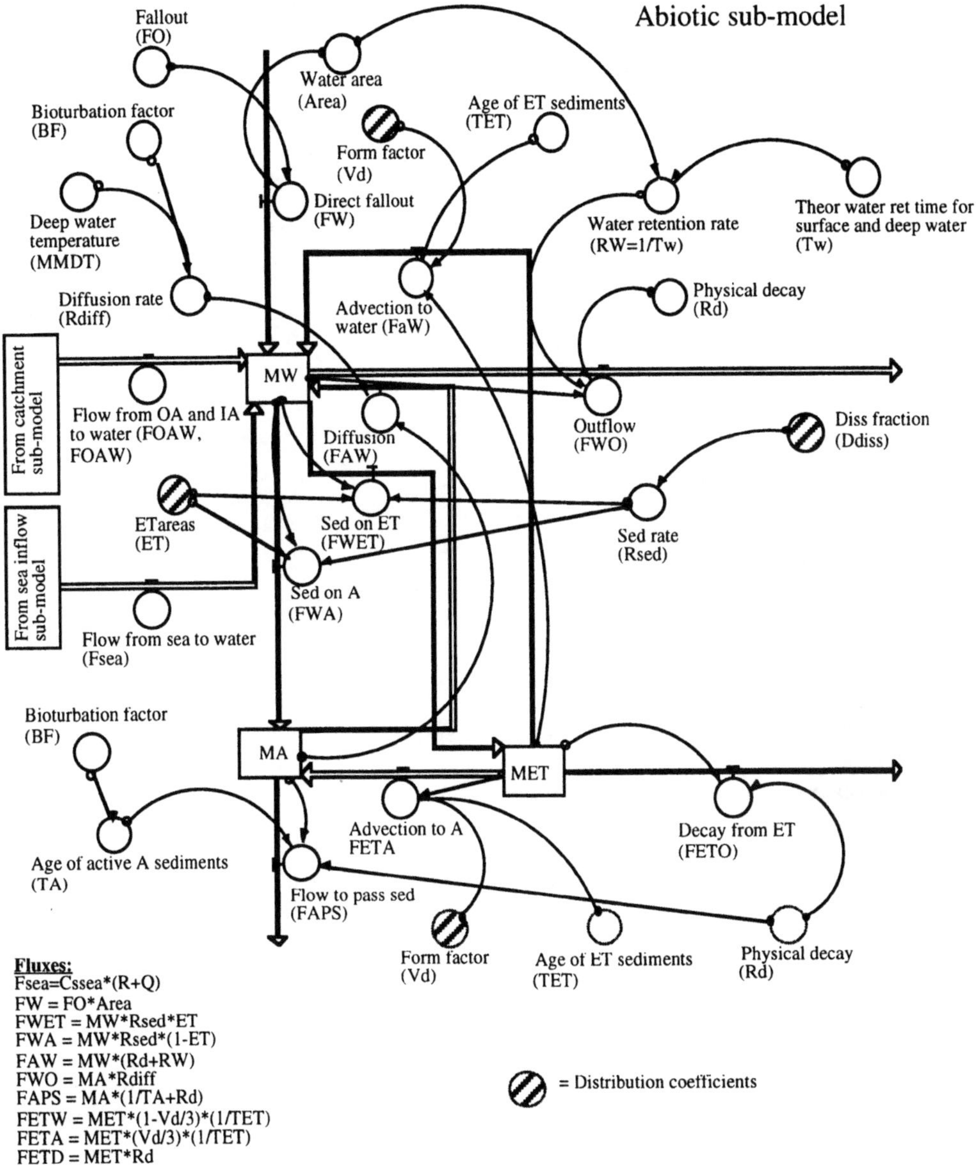

Fig. 2.47. A summary of the abiotic part of the model.

The differential equation for the concentration (i.e., C_F in Bq ^{137}Cs per kg wet muscle weight of fish) is the same (see eq. 1.46).

The biouptake, however, is different and for coastal areas and it is given by:

$$F_{IF} = Y_{SalB} \cdot Y_{AU} \cdot Y_K \cdot BMF \cdot SMTH(C_{diss}+C_{parT}, BUD, C_{diss}+C_{parT}) \qquad (2.25)$$

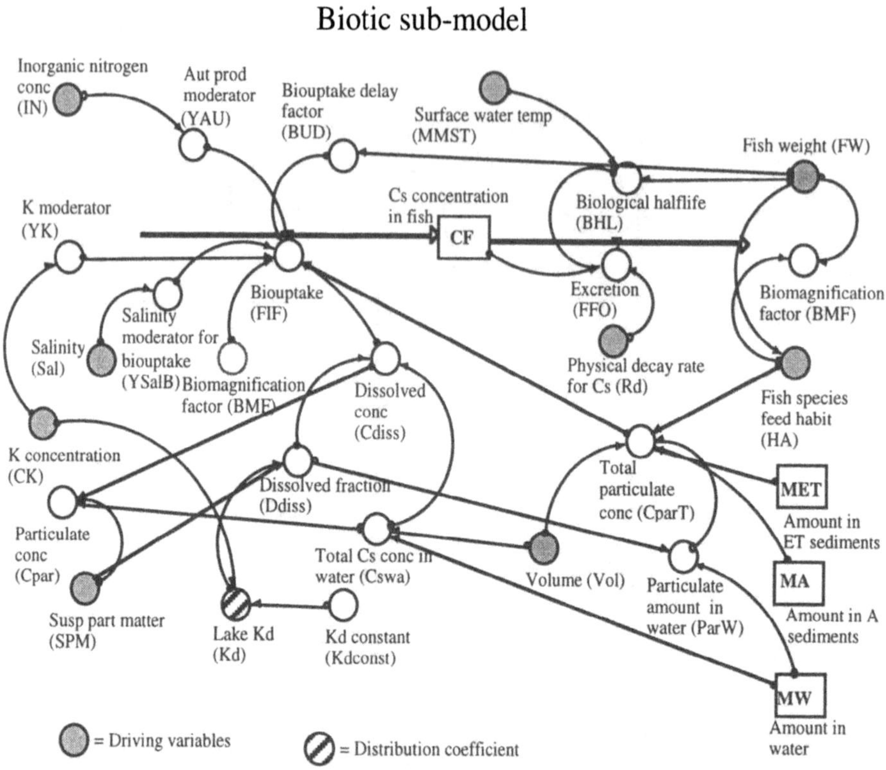

Fig. 2.48. An outline of the biotic part of the model.

where

Y_{SalB} = a dimensionless moderator describing how water salinity influences biouptake of radiocesium;

Y_{AU} = a dimensionless moderator for autochthonous production ("biological dilution"; see Håkanson and Peters, 1995); note that this moderator is NOT the same as that given by eq. 1.49 for lakes;

Y_K = a dimensionless potassium moderator (given by eq. 1.50);

BMF = a biomagnification factor;

SMTH = a smoothing function (see Figure 1.13 for illustration and definition of a smoothing function) based on the concentration of radiocesium in the dissolved phase, C_{diss}, and in the total particulate phase, C_{parT}, and a biouptake delay factor (BUD).

These factors affecting the biouptake will now be examined.

The moderator for autochthonous production, Y_{AU}, is meant to account for the principle of "biological dilution". That is, a given load of a radionuclide or a metal to a coastal area will be distributed over a larger biomass in a coast with higher bioproduction than in a coast with a lower bioproduction. Nitrogen is often the limiting nutrient in coastal areas and the Y_{AU}-value is meant to account for this in a simple manner. It is based on the most bioavailable nitrogen form, inorganic nitrogen (IN). This model uses the same approach as the lake model (see Håkanson, 1999) for radiocesium, which gives excellent predictions for cesium in lake fish. We have:

- If IN > 20 (μg/l), then $Y_{AU} = 8$
- If IN ≤ 20 (μg/l), then $Y_{AU} = 28\text{-IN}$ (2.26)

If the coastal area is mesotrophic (i.e., if the IN-concentration is higher than about 20 μg/l, see Table 2.5), it is less important to account for this principle of "biological dilution " than if the coast has a lower trophic level. This moderator accounts for this principle in a linear way. This will improve predictions only if the empirical value of the IN-concentration used is reliable. It should be noted that the moderator will cause an increase in biouptake by a factor of 2 if the IN-concentration is low (= 12, which is typical of oligo- to mesotrophic coastal waters), since the moderator gives a value of 8 if IN is higher than 20, and a value of 16 if IN is 12. The basic explanation of this moderator, and of the fact that a similar moderator increases the predictive success for radiocesium in oligotrophic lakes, is that if the biomass is very small, then a given load of radiocesium is distributed over a small biomass and the Cs-concentration (Bq per kg biomass) must by definition become higher.

Figure 2.49 illustrates how different IN-concentrations, and hence different levels of bioproduction, influence the predicted Cs-concentration in fish, here 1000 g ww piscivores and 10 g ww planktivores. The higher the bioproduction, the more pronounced the biological dilution and the lower the Cs-concentrations in fish. This figure also gives simulations for one of the very few feasible remedial measures for radionuclide contamination of coastal areas, fertilisation. The concentration of nutrients in coastal areas can be influenced by emissions from different sources, for example from point-source emissions like fish farms and water purification plants. Generally, one would need a mass-balance model to predict realistic changes in nutrient concentrations related to given emissions or remedial measures. Since the water retention time of coastal areas is often 2-10 days, it is evidently difficult to maintain high concentrations of nutrients in coastal areas, and, hence, difficult to use fertilisation as an effective remedial measure for radionuclide concentrations in water and biota. Tables 2.7 and 2.8 give typical values for chemical and biological water variables in coastal areas. Increasing the concentration of inorganic nitrogen by a factor of 2 also means probable changes in Secchi depth, gross sedimentation and the amount of suspended particulate matter. If realistic assumptions about possible changes in nutrient concentrations are made, then these simulations indicate that this is NOT a relevant remedial measure, first because it would not be very efficient, secondly because it would not be acceptable for ecological reasons to fertilise coastal areas, since many European coastal areas are already severely eutrophicated, and thirdly because it would be very costly to try to maintain the higher nutrient concentrations.

The potassium moderator, Y_K, expresses (eq. 1.50) the influence of the K-concentration on the biouptake for fish. Note that this moderator may be used for all types of cells (including plant cells).

The biomagnification factor (BMF) is the same for lakes and coastal areas. If the fish weight (WF) is < 10 g ww, then BMF = 25 for all types of fish; if WF > 10 g ww, then 250/WF for planktivores, 2 for benthivores, 3 for omnivores and 4 for piscivores.

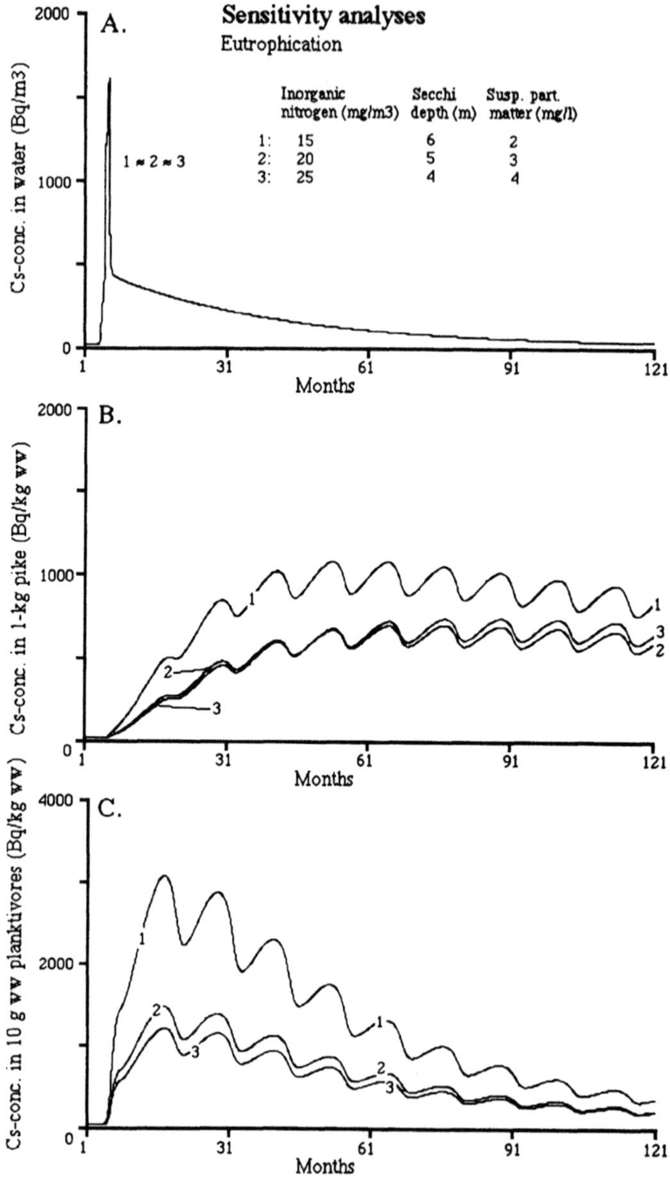

Fig. 2.49. Sensitivity test for the IN-concentration for values for total Cs-concentrations in water (A) and Cs-concentrations in 1000 g ww piscivores (B) and Cs-concentrations in 10 g ww planktivores (C).

The biouptake delay factor (BUD) is the same for the lake model and the coastal model (see eq. 1.51). It accounts for the fact that large/old animals high up in the nutrient chain do not respond as quickly as small/young animals to changes in concentrations in water.

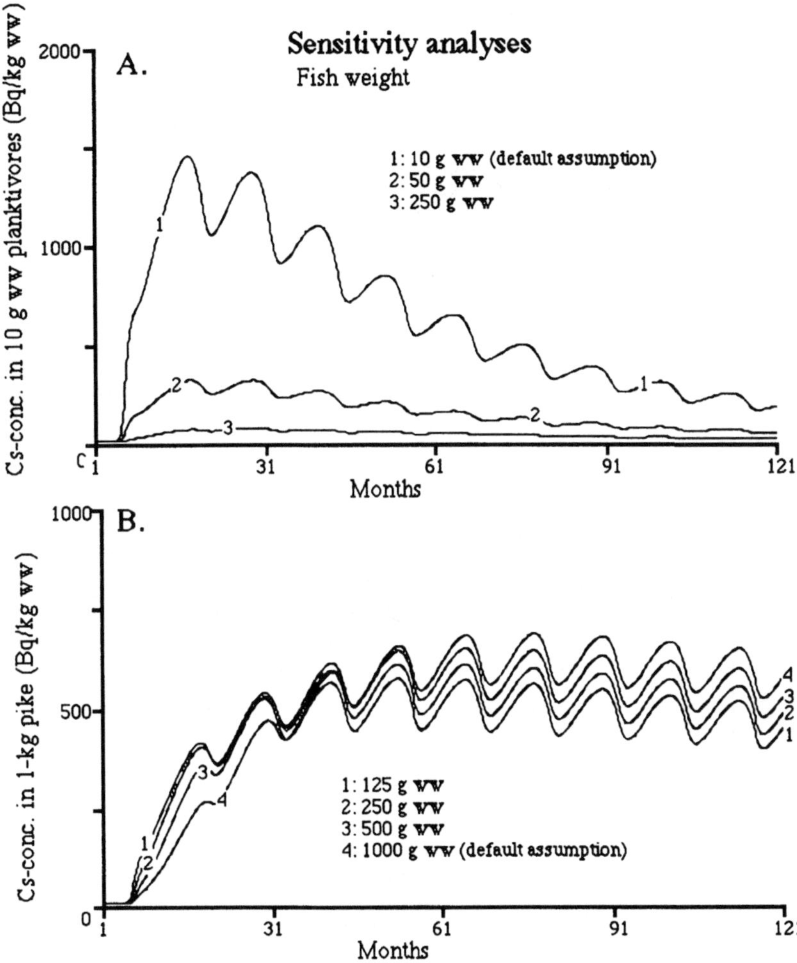

Fig. 2.50. Sensitivity test for the fish weight for values for Cs-concentrations in 10 g ww planktivores (A) and 1000 g ww piscivores (B).

Figure 2.50 illustrates how the biouptake delay factor (BUD) actually works for 10 g ww planktivores and 1-kg ww piscivores. This figures give sensitivity analyses for different weights of fish. The default values (10 g ww for planktivores and 1000 g ww for piscivores) have been varied while all else was kept constant. The larger the fish, the slower the biouptake, the longer the biological halflife in fish, and the lower the peak concentration in planktivores. The biological halflife also depends on the water temperature and this will be demonstrated in a subsequent sensitivity test.

Thus, the weight of the fish is important. Table 1.8 gave a summary of information (Brittain, 1988) regarding common European fish species from coastal/marine environments, typical weight ranges and feeding habits. This information is used to identify feeding habits and typical weights for all species of fish discussed in this context.

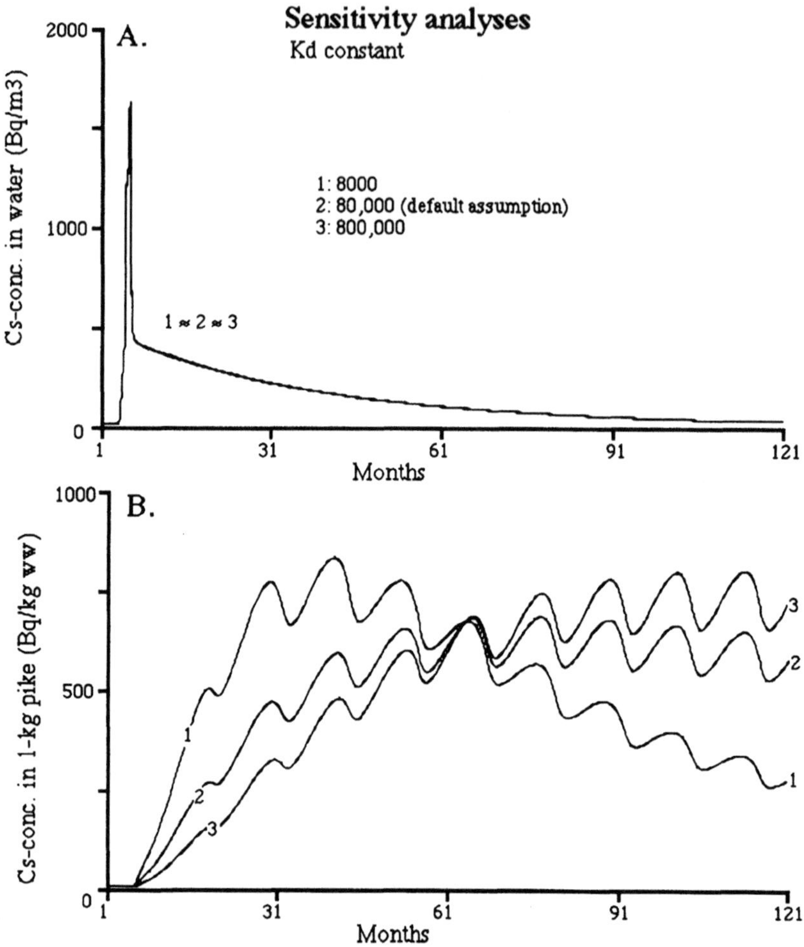

Fig. 2.51. Sensitivity test for the Kd-constant for values for Cs-concentrations in 10 g ww planktivores (A) and 1000 g ww piscivores (B).

The concentration of radiocesium is the dissolved phase, C_{diss} is calculated using eq. 1.53. Note that C_{parT} is the total concentration of radiocesium in (1) particulate phases suspended in the water plus (2) the amount of ^{137}Cs in active A-sediments and ET-sediments.

The default value used in the basic expression for Kd (i.e., the constant 800,000) and the reliability of the characteristic K-concentration greatly influences the Cs-concentration in biota. Figure 2.51 gives sensitivity tests for the Kd-constant. The default value of the Kd-constant is set to 800,000 l/kg. There are, however, uncertainties about this value, and another approach is to express Kd as a function of the K-concentration. The values for the Kd-constant have been varied over a wide range, from 0.1 times the default value to 10 times the default value. This has no clear effect on the Cs-concentration in water but certainly influences all predictions for fish, as illustrated by these sensitivity tests for Cs-concentrations in 1-kg piscivores. Kd influences the

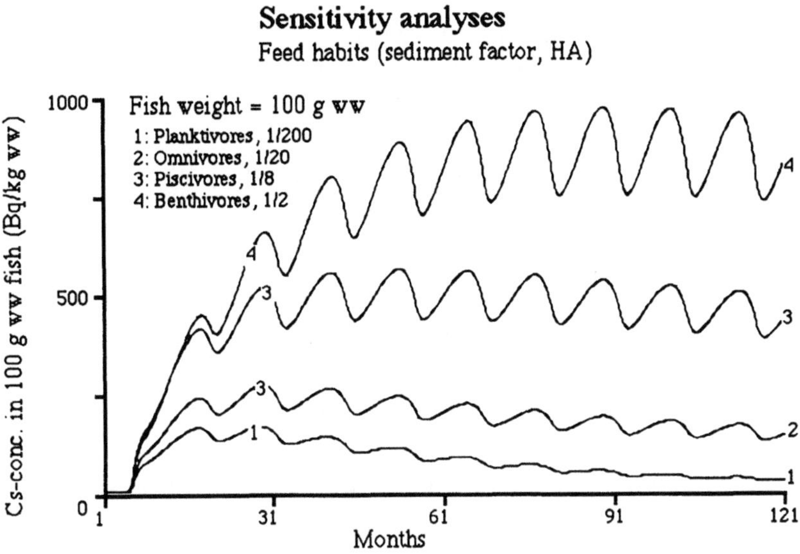

Fig. 2.52. Sensitivity tests for the feeding habit (the sediment factor, HA) for Cs-concentrations in 100 g ww fish.

dissolved fraction (and hence also the particulate fraction) and also one key abiotic variable, the retention of ^{137}Cs in the coastal system. To set the proper value, or derive a proper sub-model for the Kd-constant for coastal waters, is one important task for the calibrations of this model.

In this model also, HA means feeding habit. The following default values are used for HA: benthivores 1/2, piscivores 1/8, omnivores 1/20 and planktivores 1/200.

The expression for the biological halflife (BHL) comes from Rowan and Rasmussen (1995), see equations 1.56 and 1.57. The BHL-value depends on the weight of the fish (WF) the surface water temperature (MMST in °C; note that the term epilimnetic is not used for coastal water, only for lakes).

There are, as stated before, three ways to get temperature data:

(1) From empirical measurements in the coastal area. This is generally the most desirble approach. It is, however, evident that adequate sets of reliable empirical data are not available from many coastal areas.

(2) From standard meteorological tables. Default values used for MMST for Nordic environments in this approach are:

Month:	1	2	3	4	5	6	7	8	9	10	11	12
MMST:	0	0	3.0	4.0	7.0	11.0	16.0	20.0	15.0	7.0	4.0	1.0

These are meant to be typical mean monthly water temperatures in °C.

(3) From predictive sub-models (see Ottosson and Abrahamsson, 1998).

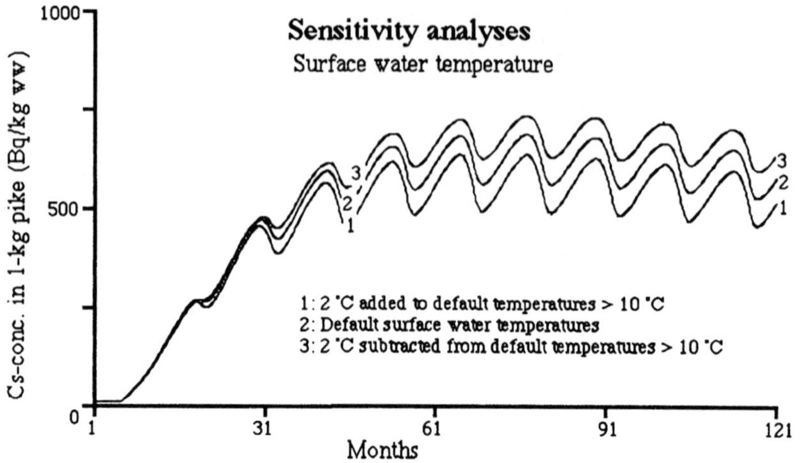

Fig. 2.53. Sensitivity test for Cs-concentrations in 1000 g ww piscivores at different surface water temperatures.

Figure 2.52 gives sensitivity analyses for different feeding habits (HA) if the fish weight is set to 100 g ww. The Cs-concentrations for fish of this weight are, and should be, highest for benthivores and piscivores, lower for omnivores and lowest for planktivores. The two reasons are the biomagnification factor (highest for piscivores and lowest for planktivores) and the influence from the sediments (HA, highest for benthivores and lowest for planktivores).

Figure 2.53 gives sensitivity tests for how surface water temperatures influence Cs-concentrations in 1000 g ww piscivores. The Cs-concentrations in fish are highest at lowest temperatures because of lower metabolic rates and longer biological halflives. It is, however, not as simple as this, as the temperature also influences stratification, mixing and the volumes of the surface and the deep water. The model predictions in Figure 2.53 give the results for the default coastal area when 2°C was added and subtracted, respectively, to the default temperatures during the summer period (i.e., for temperatures > 10°C).

2.3.3. PANEL OF DRIVING VARIABLES

This panel (see Figure 2.54) is divided into two main parts, the area-specific variables, which must be available for every coastal area (these are the obligatory driving variables), and the model variables (the model constants, which should NOT be changed for different coasts but should be regarded as generic model values). The figure also gives priority classes for the driving variables: (1) obligatory driving variables, (2) desired driving variables and (3) requested driving variables, which contribute relatively little to the model uncertainty and the given default values (see Table 1.2) would generally be sufficient for model predictions.

The area-specific variables are divided into:

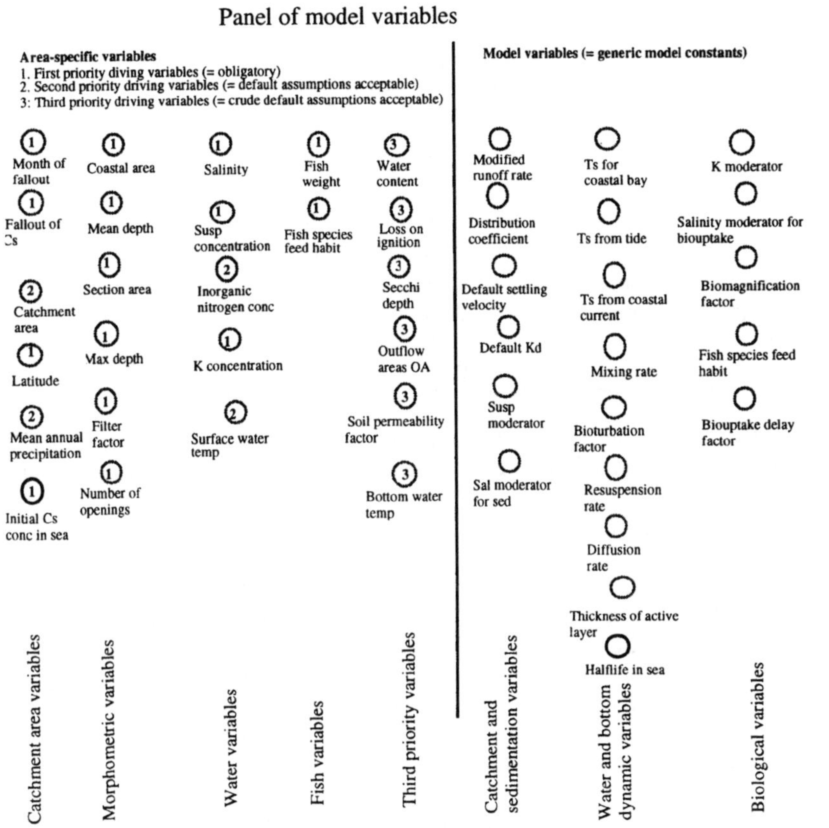

Fig. 2.54. Panel of driving variables.

- catchment variables: fallout, initial Cs-concentration in the sea (related to fallout), latitude, catchment area, outflow areas, soil type (or soil permeability factor) and mean annual precipitation;
- morphometric variables: coastal area, mean depth, section area, maximum depth, filter factor and number of openings;
- water variables: salinity, concentration of suspended particulate matter, concentration of inorganic nitrogen, K-concentration, Secchi depth, monthly mean surface water temperatures and deep water temperatures;
- sedimentological variables: water content of active A-sediments, organic content of active A-sediments (= loss on ignition).

It is generally sufficient to determine the climatological variables (surface and deep water temperatures and mean annual precipitation) for larger regions, so they may not have to be determined for each coastal area. This means that they can be regarded either as area-specific variables or as model variables.

Table 2.9. Average number of samples in individual coastal areas (n), coefficients of variation (CV = SD/MV) and the calculated statistical error (L) at 95% confidence level for different water quality and sediment trap variables (based on Wallin et al., 1992).

Variable	n	CV	L (%)
Secchi depth	12	0.19	11.4
Chlorophyll	7	0.25	20.3
Total-N	7	0.13	10.3
Inorganic-N	7	0.31	25.0
Total-P	7	0.16	13.0
Inorganic-P	7	0.28	22.6
Sedimentation (surface traps)	5	0.58	46.1
Sedimentation (bottom traps)	5	0.50	39.8
O_2-concentration	12	0.26	15.1
O_2-saturation	12	0.25	14.8

Characteristic CV-values for many of the model variables are given in Table 2.9 (see also Table 1.2). Many of these CV-values will be used later in sensitivity and uncertainty analyses.

The model variables are categorised into:

- abiotic variables: fall velocity (v_{Cs}), age of A-sediments (T_A), age of ET-sediments (T_{ET}), distribution coefficient between inflow and outflow areas (D_{OAL}), the initial retention rate of radiocesium in outflow areas (R_{Rd}), the diffusion rate (R_{Diff}) and the halflife of radiocesium after fallout in the sea outside the given coastal area (T_{sea}).
- biotic variables: feeding habit (sediment factor, HA), biomagnification factor (BMF), K-moderator (Y_K), autochthonous moderator (Y_{AU}), biouptake delay factor (BUD) and salinity moderator for biouptake (Y_{SalB}).

To make a prediction, one must also decide which fish species are to be predicted and the mean population weight of the selected fish species.

2.3.4. INVENTORIES

A critical question in most calculations dealing with fluxes, uncertainties in fluxes and ranking of fluxes concerns the inventory problem, i.e., where is the radiocesium stored?

Figure 2.55 gives the inventory (Bq) of radiocesium after 120 months in inflow areas, outflow areas, A-areas, ET-areas and in coastal water under default conditions (fallout = 50,000 Bq/month during month 5) with the catchment sub-model and the given default inflow from the sea. Note that there is a 3-4 order of magnitude difference between the amount in the water and in inflow areas and that the amounts in ET-areas are reduced much more rapidly than the amounts in A-areas.

A similar inventory for 200 years is given in Figure 2.56. It is interesting to note the marked difference in the impression of the recovery process if the scale of the y-axis is based on absolute data or logarithmic data, and that the amounts in the water, ET-areas and A-areas are actually reduced relatively quickly while the amounts in inflow areas and outflow areas of the catchment are reduced more slowly.

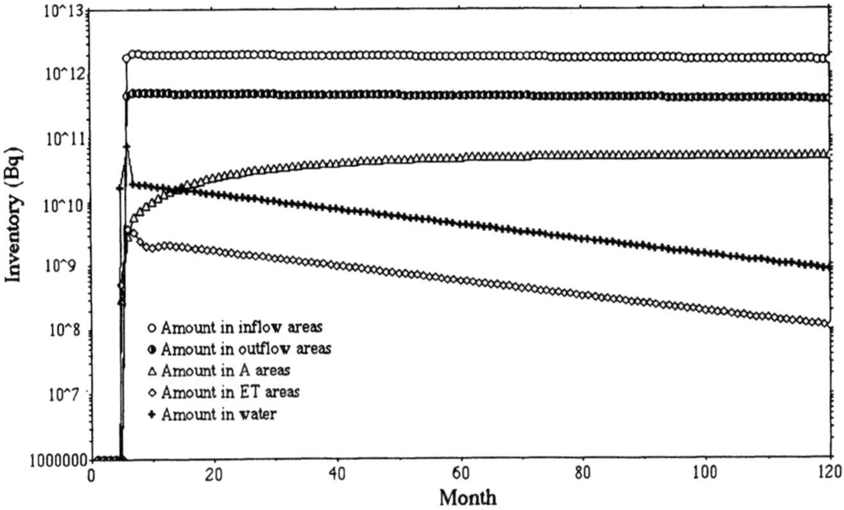

Fig. 2.55. Inventory, months after fallout.

2.3.5. REMEDIAL MEASURES

Various realistic remedial measures for reducing radiocesium concentrations in water and biota have been tested and discussed in many contexts (see IAEA, 1999; Håkanson, 1999). The aim here is not to repeat such discussions and results. It is evident from the fertilisation simulation presented in Figure 2.49 that coastal area fertilisation is NOT generally a realistic remedial measure. In this section, the model will be used to address two specific aspects of remediations: first a scenario in which dredging is tested as a way of removing contaminated sediments, and second, a "worst-case" scenario in which remedial measures would be needed the most.

2.3.5.1. Dredging

Figure 2.57 gives results from a test in which 50% of the radiocesium in the active A-sediments was removed at given months, i.e., the fallout month, 5, or months 6, 7, 8 or 12. Note that it is NOT realistic to assume that 50% of the inventory in A-sediments can be removed, and it is certainly not realistic to assume that this can be done the same month as the fallout, especially not for large coastal areas. Thus, these simulations are meant to give indications as to whether dredging in coastal areas is advisable, and also to provide realistic expectations for such endeavours. The results are given for piscivores and benthivores (which would be expected to respond more to this treatment than planktivores). From the results given in Figure 2.57, it should be noted that this would probably be a costly and inefficient remedy. Just like fertilisation, it CANNOT be generally recommended in coastal areas. This means that actions like fish bans and domestic food treatment, e.g., salting of fish before consumption (see IAEA, 1999),

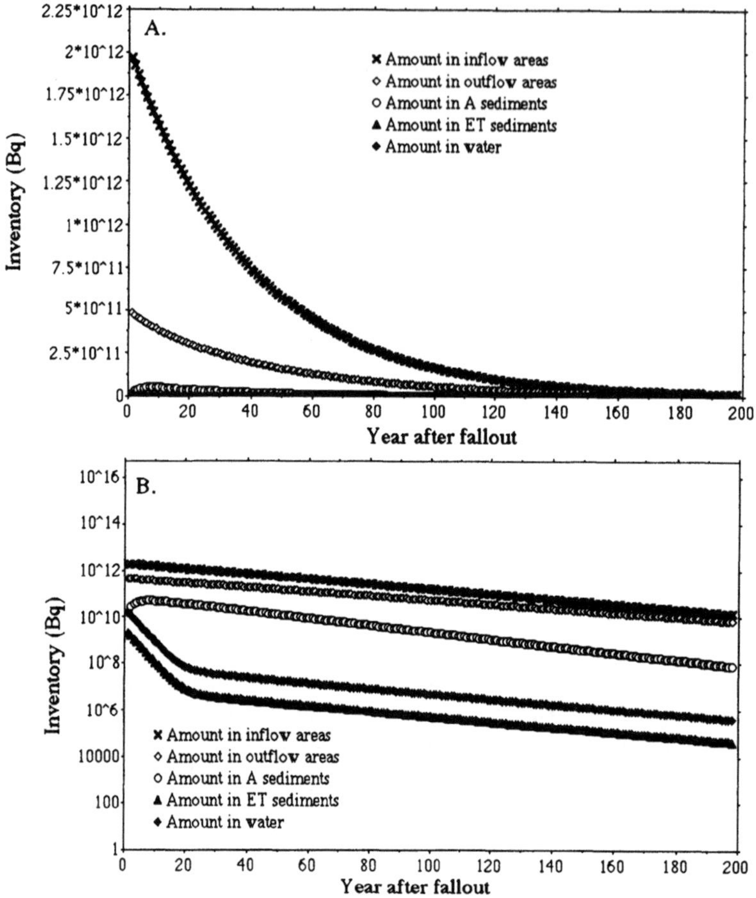

Fig. 2.56. Inventory, years after fallout.

should be carried out, but that these remedial measures in coastal areas CANNOT be recommended.

2.3.5.2. Worst-case scenario

The aim of the test in Figure 2.58 is to illustrate a "worst-case" scenario. The presuppositions are given in the figure. The fallout is 100,000 Bq/m^2 (twice the default value), and many of the factors influencing the "peak and the tail" have been set to

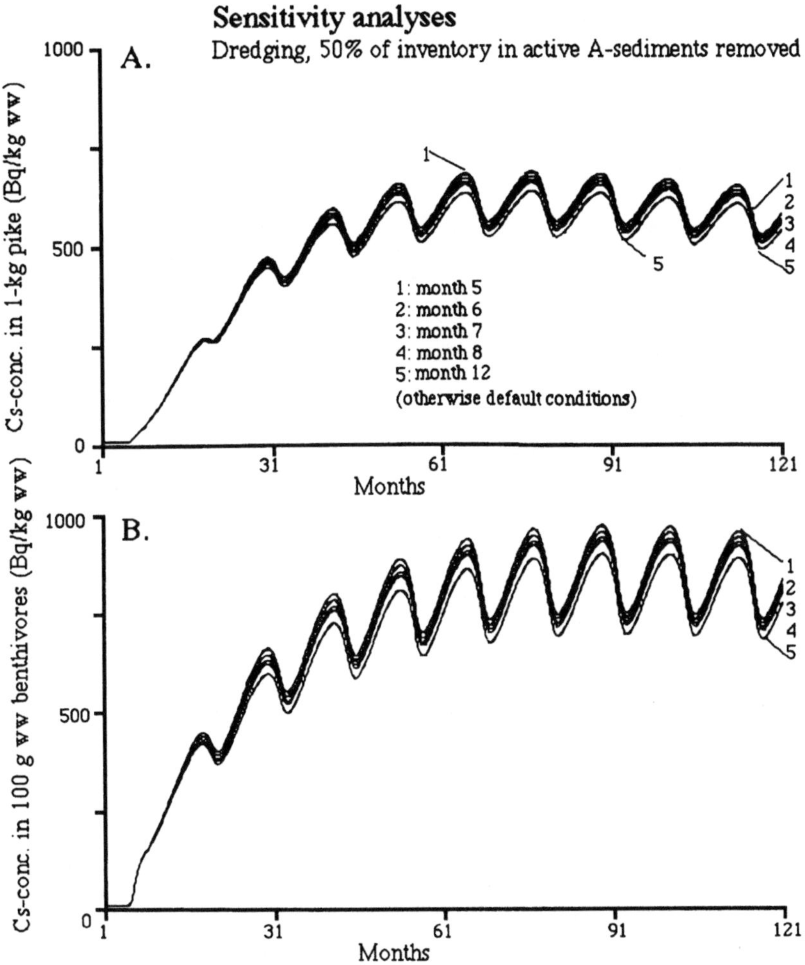

Fig. 2.57. Dredging as a remedial measure. In this scenario 50% of the inventory in A-sediments is removed in different months after the fallout.

maximise the peak and prolong the tail. The area is large, enclosed and has a small filter factor; the inflows from the catchment area and the sea are accounted for and the calculations concern very large benthivores. From this simulation, it should be noted that it takes about three years until the guideline value of 1500 Bq/kg ww for consumption and marketing of fish is reached and about 10 years until the maximum Cs-concentration is reached (max. value about 2200 Bq/kg ww) in these animals. The main reason for this slow response is that considerable amounts of radiocesium are stored in the active A-sediments. The curves for the Cs-concentrations in active A-sediments and large benthivores are rather parallel, except for the seasonal variations in Cs-concentrations in the animals, which depend on the metabolic activities related to water temperature.

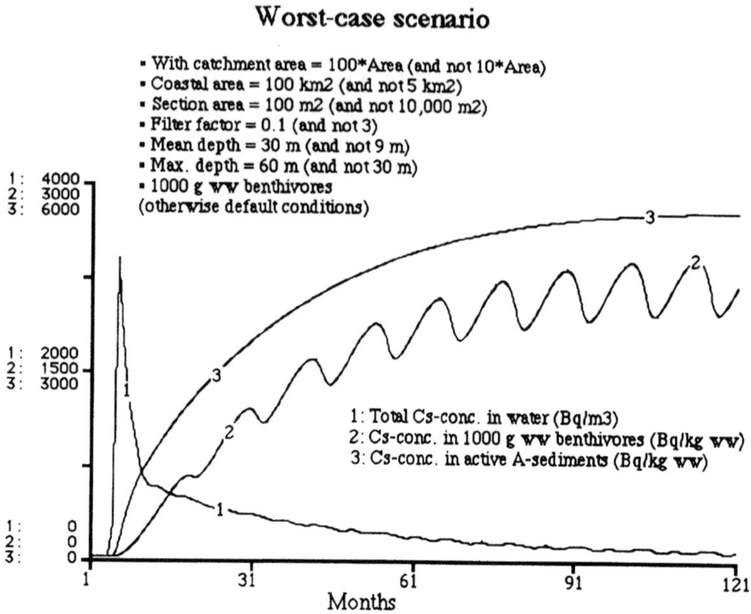

Fig. 2.58. The worst-case scenario.

2.4. Critical Model Tests

The aim of the following section is to present critical model tests according to the methods already discussed for lakes. These tests are very important since the overall model has not been calibrated or validated, unlike the different sub-models.

2.4.1. SENSITIVITY AND UNCERTAINTY ANALYSES

There are two targets for the following tests, i.e., Cs-concentrations in water and in fish. Generally, in dynamic models like this radiocesium model, it is more difficult to predict concentrations of toxic substances in biological compartments than in abiotic compartments (see Håkanson and Peters, 1995). Thus, one should expect that the sensitivity and uncertainty tests will demonstrate a smaller CV for the predicted Cs-concentrations in water compared to fish. These tests have focused on the model variables, i.e., on the most uncertain parts of the model - the rates and distribution coefficients which have the highest CV-values (as given in Table 1.2). The following model variables have been included in the tests:

1. All factors influencing biouptake (Bioupt). CV = 0.35.

a. The weight of the fish. There is always a certain uncertainty concerning which mean population weight should be used. Generally, however, there would be empirical data for this. Such empirical data will give a mean weight and a CV for fish weight. This particular uncertainty evidently cannot influence the predictions for the Cs-concentrations in water, but it may be important for the predicted Cs-concentrations in fish. The idea with these tests is to see how the overall uncertainty in biouptake influences the Cs-predictions for fish; and the individual

factors influencing this uncertainty, like the weight of the fish which influenced the biouptake delay factor, iare not tested. In the following tests, we will use a CV of 0.35 for the biouptake. This means that if the biouptake rate is 1, all the uncertainties associated with biouptake should create a variability from 0.3 to 1.7 with a 95% certainty (corresponding to 2·SD).
b. The biomagnification factor (BMF). The approach used in this model emanates from rather extensive empirical tests (see Rowan and Rasmussen 1994, 1995). This uncertainty only affects the Cs-concentration in fish and is also included in Bioupt.
c. The biouptake delay factor (BUD) influences the biouptake of radiocesium in such a manner that fish larger than 100 g ww would be expected to have a slower biouptake than smaller fish, which are often planktivores and reach a dynamic steady-state within one month (the calculation time of the model). This is also an uncertain model variable. This uncertainty only affects the Cs-concentration in fish and is included in Bioupt.
d. The potassium moderator (Y_K). This moderator is based on fundamental mechanistic processes regulating the competition between potassium and cesium uptake via cell membranes. The uncertainty for this moderator only affects the Cs-concentration in the fish and is also included in Bioupt.
e. The autochthonous moderator (Y_{AU}), which accounts for "biological dilution" and is related to the IN-concentration. This factor is relatively uncertain, although the underlying effect of "biological dilution" is well documented and almost self-evident since the definition of the mean Cs-concentration in fish is amount of radiocesium in fish divided by the fish biomass. This uncertainty only affects the Cs-concentration in fish and is included in Bioupt.
f. The salinity moderator (Y_{SalB}), which accounts for the influence of salinity on biouptake. This uncertainty only affects the Cs-concentration in fish.
g. The feeding habit (or sediment factor, HA), reflecting the exposure from all particulate phases (benthic pathways, etc.). This factor is large for benthivores (1/2), smaller for piscivores (1/8) and omnivores (1/20) and very small for planktivores (1/200). The actual values for the different exposures to sediments and particulate matter are, however, uncertain. This uncertainty only affects the Cs-concentration in fish and is also included in Bioupt.

2. All factors influencing excretion (Excret). CV = 0.25.
a. The biological halflife (BHL). The approach used in this model has been tested in several contexts (see Rowan and Rasmussen 1995; Håkanson et al., 1996a, b; IAEA, 1999). It is fairly well known that BHL depends on the weight of the fish and the water temperature. This uncertainty only affects the Cs-concentration in fish and is also included in Excret.
b. The surface water temperature (MMST). Even if very reliable data for the mean monthly surface water temperatures are available from long measurement series, such data may not be correct for future temperatures. Information on mean monthly surface water temperatures is generally available from Hydrological and meteorological tables or from models. This uncertainty only affects the Cs-concentration in fish and is included in Excret.

3. All factors influencing sedimentation (v). CV = 0.5.
a. The partition coefficient for cesium in water (Kd, or rather the Kd-constant 800,000). The model for Kd is based on mechanistic processes describing how K-ions influence the binding of Cs-ions to carrier particles. There are, however, uncertainties concerning the constant 800,000 in this sub-model.
b. The fall velocity (v_{Cs}). The value of the fall velocity is, together with the approach for determining Kd, the key regulator of the sedimentation process, and hence the amount of radiocesium in sediments, and the internal loading. The v_{Cs}-value is derived from circumstantial information and not from direct measurements.
c. The concentration of suspended particulate matter (SPM) in the water, which influences the settling velocity of particulate cesium, the Kd-value, the deposition of matter, and hence the age of the sediments, and the retention time of radiocesium in sediments. The values of SPM can be measured and/or predicted quite well.

4. The transport from land to water (Q). CV = 0.25.
a. The fraction of outflow areas (OA), which is used as a distribution coefficient. The value can be determined quite well from standard maps showing bogs, mires and upstream lakes. This uncertainty can influence the secondary transport of radiocesium from land to water, and hence Cs-concentrations in water and fish.
b. The distribution coefficient regulating the transport from inflow area directly to the water and/or to outflow areas (D_{OAW}).
c. The basic runoff rate from outflow areas (R_{Rd}). The initial value of 0.04 (i.e., a 4% loss during the first year after the fallout) is derived from several studies (see Håkanson et al., 1988a; IAEA, 1999). These studies have also shown that this value is uncertain because many factors can influence the value, e.g., soil type, precipitation, season of the year and ice-conditions.
d. The soil type or the soil permeability factor (SPF). This is an uncertain factor and there are major differences in soil types within and among catchment areas (soils always appear in a patchy manner, even within 1 m^2), so it is generally difficult to assess what value should be used to characterise the entire catchment area.

5. Resuspension (Resusp). CV = 0.5.
a. The age of radiocesium on ET-areas (T_{ET}). If this age is set too short, say 1 day, the ET-areas would function as E-areas. If, on the other hand, the age is set too high, the ET-areas would function as A-areas. What value should be used? Generally, it is about 1 year for lakes and about 1 month for coastal areas, but this is an uncertain value.
b. The age of radiocesium on A-areas (T_A). This is determined by the sedimentation of matter, the definition of the thickness of the active layer and the bioturbation, all of which are uncertain factors.
c. The ET-area (ET). This is a distribution coefficient in this model separating sedimentation of radiocesium on ET- and A-areas. The ET-areas can be measured and modelled quite well.

6. Diffusion (Diff). CV = 0.5.
The diffusion of radiocesium from A-areas (Diff). There are, as mentioned, some empirical data suggesting that a general default value for the diffusion rate can be set to 0.002. This value depends on the sedimentation of material, and on the deep water temperature and bioturbation.

7. The theoretical surface water retention time (Ts). CV = 0.25.
This value is important. It is calculated from the inflow of water from the catchment and from the sea and it regulates the outflow of radiocesium from the coast (the openness of the "exit gate"). Ts is calculated from several rather uncertain models.

8. The theoretical deep water retention time (Td). CV = 0.35.
This value is not generally very important. It is calculated either from a morphometric regression or is related to Ts.

9. The Cs-concentration in the sea and/or adjacent coastal areas (Cs_{sea}). CV = 0.5.
The approach for estimating Cs_{sea} in this model is crude and uncertain. Basically, one would need a very extensive model for the entire ocean to calculate this value.

10. The ecological halflife for radiocesium in the sea outside the given coastal area and/or in adjacent coastal areas, T_{sea}. CV = 0.5. As for Cs_{sea}, one would need a very extensive model to predict T_{sea}.

11. The transport from active to passive A-sediments (Pass). CV = 0.5.
This value depends on several uncertain considerations regarding the depth of the active layer, gross sedimentation, sediment water content, sediment organic content (= loss on ignition) and, hence, sediment age.

12. Last but not least, the fallout (Fall). CV = 0.1.
It is evident that in spite of the fact that the fallout can be measured quite well, the value used for the fallout onto the coastal area, the catchment area and the area outside the given coast can be quite uncertain, especially for large catchments. This is the primary driving force for all subsequent calculations. The CV of the fallout has been set to 0.1.

The aim of the following tests is to produce a ranking of these uncertainties for the two target variables. It is evident that all uncertainties cannot be of equal importance in a given coastal area and that the ranking must differ among coastal areas

All of the following tests use the data for the default coastal area listed in Table 2.3. All calculations use 100 runs and all uncertainties are given for month 11, i.e., 6

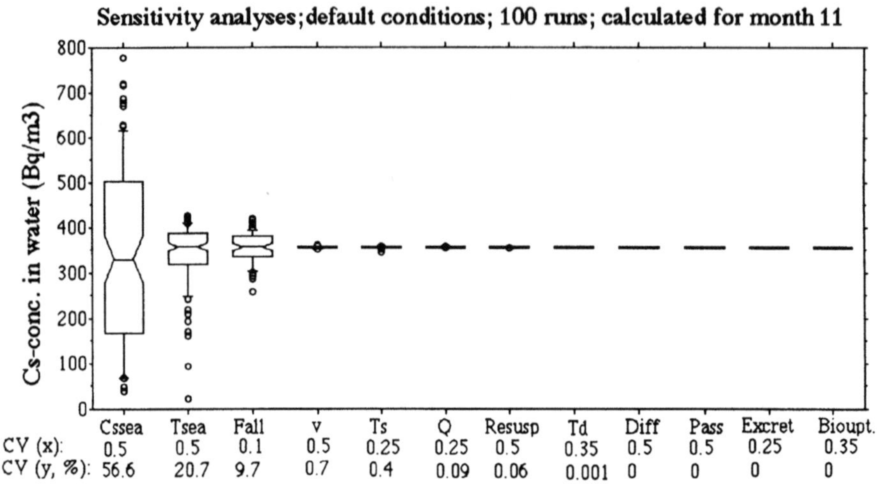

Fig. 2.59. Sensitivity tests calculated for month 11 (6 months after the fallout) for the Cs-concentration in water. 100 runs. The box-and-whisker plots give median values (M_{50}), quartile values (M_{25} and M_{75}), percentiles (M_{10} and M_{90}) and outliers.

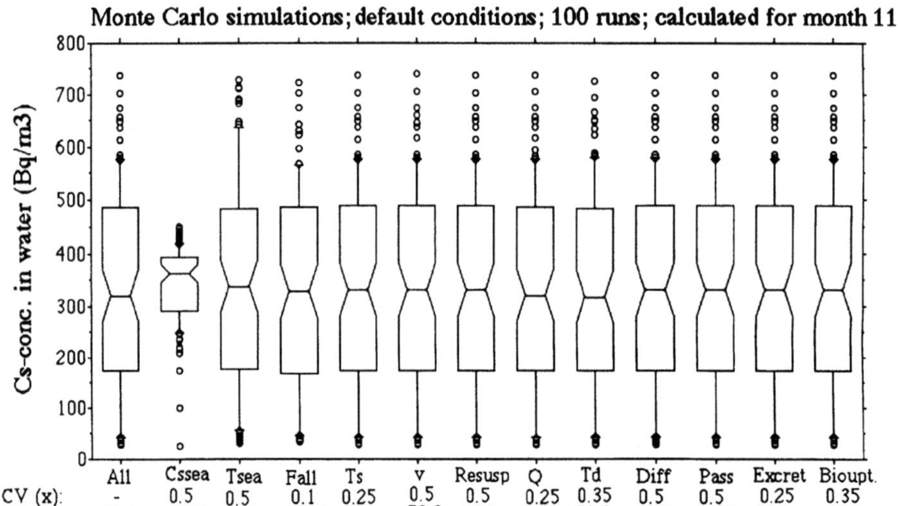

Fig. 2.60. Uncertainty tests calculated for month 11 (6 months after the fallout) for the Cs-concentration in water. 100 runs. The box-and-whisker plots give median values (M_{50}), quartile values (M_{25} and M_{75}), percentiles (M_{10} and M_{90}) and outliers.

months after the fallout. These are the presuppositions and the results depend on these conditions.

Figure 2.59 first gives the sensitivity tests for Cs-concentrations in water. The ranking is given by the CV-values for the Cs-concentrations in water and the box-and-whisker plots. The most important factor, by far, for the calculated uncertainty in the Cs-concentrations in water is the Cs_{sea}-value, which regulates the flow of cesium from the sea to the coast. A criterion for a well-balanced model is that the calculated CV-values for the target variable (y) are about the same for all model variables (see Håkanson, 1999). Thus, this is certainly a poorly balanced model. The default predictions depend very much indeed on the uncertainty associated with Cs_{sea}. It should be noted that under default conditions only three factors actually influence the overall uncertainty for the model predictions of the Cs-concentrations in water, Cs_{sea}, T_{sea} and fallout (Fall). All other model variable uncertainties are of minor importance. This is interesting and suggests that the model can be simplified, but only if these results are valid for different coastal areas. Many tests like the one given in Figure 2.59 have been carried out. It can be concluded that the uncertainties associated with Cs_{sea} and T_{sea} are generally of paramount importance, and that the importance of these two factors increases with time after fallout. This means that the model can be simplified, and in the next section, a simplified version of the model will be given.

Figure 2.60 gives the corresponding uncertainty tests using Monte Carlo techniques. It should be noted that the total CV for Cs-concentrations in water is 58.9%. The CV for empirical data for within-lake variations in Cs-concentrations is 0.3 (see Table 1.2). The correspondence between the calculated CV of 0.59 for coastal areas and the empirical CV of 0.3 for lakes is quite logical. One would expect greater CV values for rivers and coastal areas with high water turnover rates than for lakes with generally much lower water turnover rates. This also means that the CV-values selected in these uncertainty analyses for the model variables produce an appropriate calculated CV in the target variable. This is an important control of the presuppositions for the test. Generally, one should apply the precautionary principle and CV-values that are too high rather than too low for the model variables so that the calculated CV for y is higher than the empirical value.

In short, this result indicates that the assumptions concerning the CV-values for the model variables are reasonable.

The results of the Monte Carlo simulations agree very well with the results from the sensitivity analyses. The most important factors are again Cs_{sea}, T_{sea} and fallout. If these uncertainties are omitted, the CV for the Cs-concentration in water decreases the most, from 58.9% to 22.4% when Cs_{sea} is omitted, to 57.2% when T_{sea} is omitted and to 58.3% when fallout is omitted. The uncertainties associated with the other model variables are of little consequence for the default coastal area, and for many other coastal areas as well. This means that future model improvements should concentrate on getting more reliable data and/or sub-models for Cs_{sea} and T_{sea}. This would be the best way to reduce the uncertainties in the model predictions. It should be stressed again that the model is meant to be used generally and this means that many of the factors and processes accounted for play a limited role in the model predictions in relation to Cs_{sea} and T_{sea}. The model can be simplified!

It should also be stressed again that even though the values for, e.g., the ages of ET- and A-sediments, the values used for the ET-areas and the diffusion rate were not important in this coastal area, they will be of greater importance in shallower coasts. The uncertainty and sensitivity analyses are only indicative of how the model predicts. The ultimate test of predictive power is NOT sensitivity and uncertainty tests, but validation.

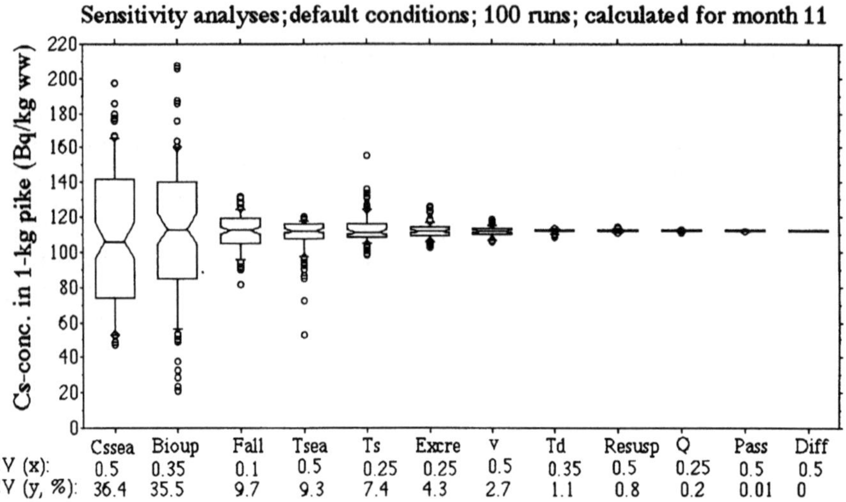

Fig. 2.61. Sensitivity tests calculated for month 11 (6 months after the fallout) for the Cs-concentration in 1-kg piscivores. 100 runs. The box-and-whisker plots give median values (M_{50}), quartile values (M_{25} and M_{75}), percentiles (M_{10} and M_{90}) and outliers.

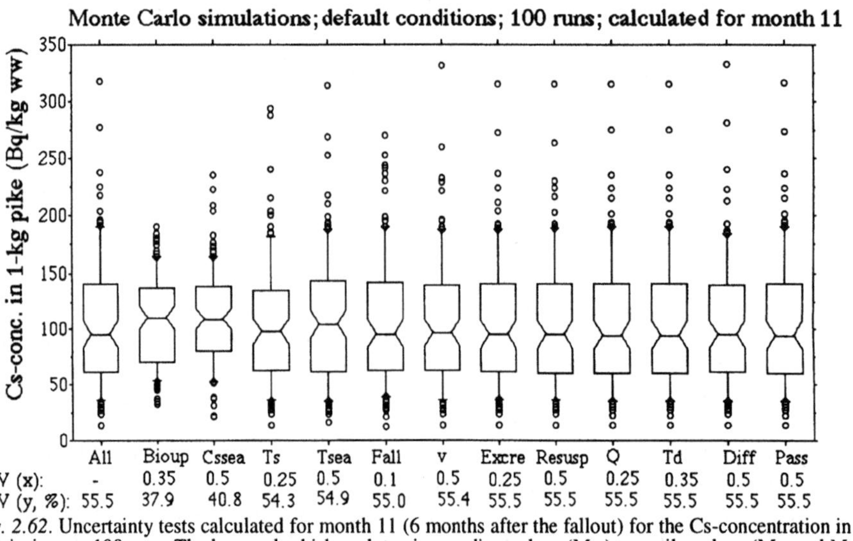

Fig. 2.62. Uncertainty tests calculated for month 11 (6 months after the fallout) for the Cs-concentration in 1-kg piscivores. 100 runs. The box-and-whisker plots give median values (M_{50}), quartile values (M_{25} and M_{75}), percentiles (M_{10} and M_{90}) and outliers.

Next, results are given for Cs-concentrations in fish (here 1000 g pike). One might expect the calculated CV-values to be higher, not because the CV-value for empirical Cs-concentrations in fish is higher, which it is not (the characteristic CV for Cs in lake fish is 0.22 and for lake water 0.3; see Table 1.2), but because it should be much more difficult to model and predict Cs-concentrations in fish, since these predictions incorporate predictions of Cs-concentrations in water as well as many other processes related to feeding habits, biouptake, etc.

The results are given in Figure 2.61 (sensitivity analyses) and Figure 2.62 (uncertainty analyses). These figures demonstrate some points outlined below:

• The calculated CV for y in the uncertainty test has not, as might be expected, increased but has decreased from 58.9 % for Cs-concentrations in water to 55.5 %. This is a small decrease and the main reason is the relatively small CV-values used for biouptake (CV = 0.35) and excretion (CV = 0.25), because the sub-models for biouptake and excretion emanate from the lake model for radiocesium, which gave excellent predictions when validated.

• The ranking of the factors contributing to this CV is very similar in the two tests. The sensitivity tests (Figure 2.61) gave:

Cs_{sea} > biouptake > fallout > T_{sea} > Ts > excretion > settling velocity; all other model variables were of little or no significance in this test.

The uncertainty tests (Figure 2.62) gave:

biouptake > Cs_{sea} > Ts > T_{sea} > fallout > settling velocity; all other model variables were of no significance in this test.

It is evident that the uncertainties associated with Cs_{sea} and T_{sea} are also key components also in the fish sub-model.

• The model is not well balanced since a few model components provide most of the model uncertainty in the predicted target variables.

2.4.2. A CONTINUOUS CONTAMINATION SCENARIO

It should be stressed once more that the assumptions and models used for the conditions in the sea outside the given coastal areas are of paramount importance for the model predictions, not just for the default coastal area, but for most coastal areas. The model predictions during the first month after the fallout depend very much on the uncertainties associated with the sub-models for the surface water retention time in the default coastal area. More generally, the biggest uncertainties are always related to the biggest fluxes. Since the fluxes between a given coastal area and the sea are generally dominant, the uncertainties associated with these fluxes are generally the most crucial ones.

The given results apply to a peak fallout of the Chernobyl type. In this section we will, as a comparison, simulate conditions related to a continuous contamination, i.e., conditions more typical of, e.g., heavy metal pollution or coastal water eutrophication.

The scenario is as follows.

1. The fallout is set to 0.

2: The concentration in the sea is set to be constant (Cs_{sea} = 100 Bq/m^3).

3. The transport to the coastal area from the catchment area is given by the Cs-concentration in the tributaries (Cs_{trib}), which is set to be 1000 Bq/m^3, i.e., 10

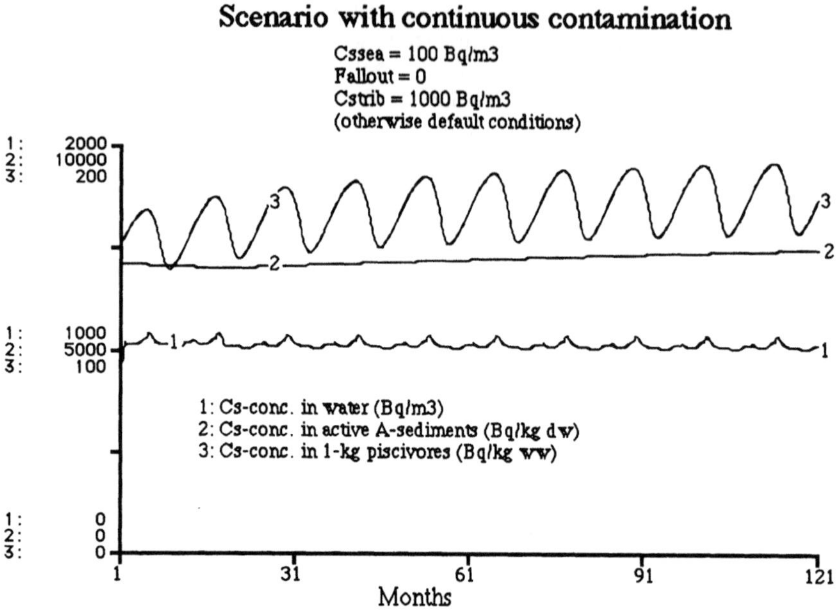

Fig. 2.63. Results from a scenario of continuous contamination.

times higher than the concentration in the sea. The idea is to see how this will influence the Cs-concentrations in the coastal water.

Otherwise the predictions are based on the default conditions, as given by Table 1.3. From these presuppositions, Figure 2.63 gives the results in terms of predicted Cs-concentrations in water (curve 1), in active A-sediments (curve 2) and in 1-kg piscivores (curve 3). It should be noted that not even this high tributary concentration would influence the Cs-concentration in the coastal water very much. The predicted value is very close to the Cs-concentration in the sea, and the small fluctuations are related to seasonal variations in tributary water discharge. That is, the conditions in the coastal area are governed by the conditions in the sea.

Figures 2.64 and 2.65 give results from sensitivity and uncertainty analyses for this scenario for Cs-concentrations in water and in 1000 g ww piscivores. In this case, it was assumed that the CV-value for Cs_{sea} was not 0.5, but 0.2. This is a typical CV-value for many coastal water variables (see Table 2.9). Such a CV-value should be based on empirical samples, i.e., it should be derived from several water samples taken at various sites outside the given coastal area during one month and calculated from the given mean value and standard deviation. The CV-value for the tributary concentration is set to 0.5, which is typical of water pollutants in rivers (see Håkanson, 1999). Only the most important model variables from the previous tests have been included in these tests, i.e., Cs_{sea}, Ts, v and biouptake (Bioupt.). There are, as can be seen, very great differences when one compares these results for continuous contamination with the results related to a pulse-type contamination.

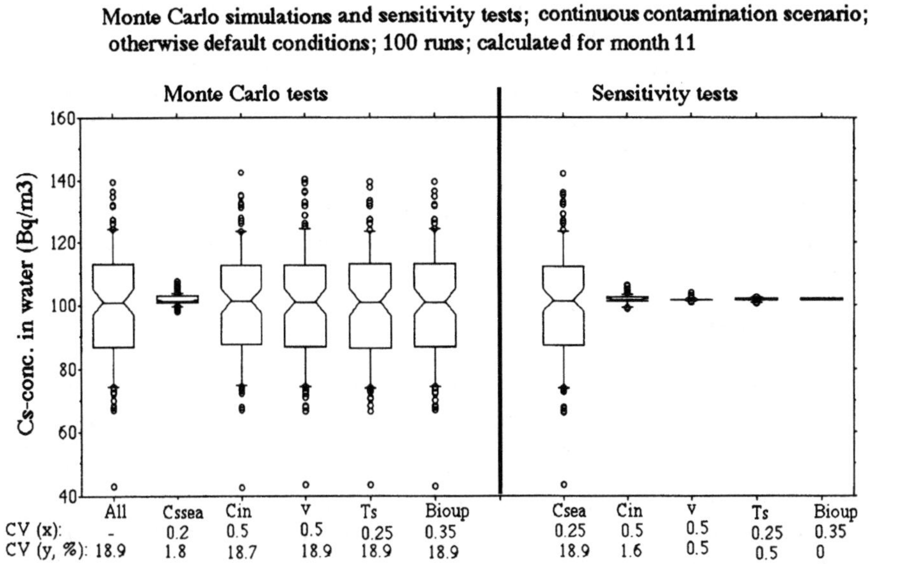

Fig. 2.64. Sensitivity and uncertainty tests calculated for month 11 (6 months after the fallout) for the Cs-concentration in water. 100 runs. The box-and-whisker plots give median values (M_{50}), quartile values (M_{25} and M_{75}), percentiles (M_{10} and M_{90}) and outliers.

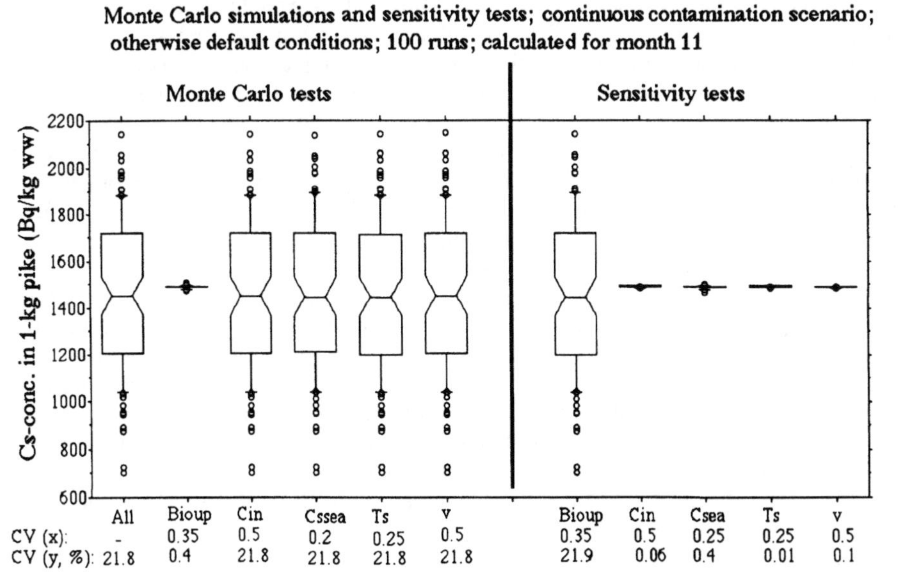

Fig. 2.65. Sensitivity and uncertainty tests calculated for month 11 (6 months after the fallout) for the Cs-concentration in 1-kg piscivores. 100 runs. The box-and-whisker plots give median values (M_{50}), quartile values (M_{25} and M_{75}), percentiles (M_{10} and M_{90}) and outliers.

• The calculated CV-values for the target variables are much lower - 18.9% for Cs-concentration in coastal water as compared to 58.9%, and 21.8% for Cs-concentrations in 1-kg piscivores, as compared to 55.5%.

• It is interesting to note also that in this case the uncertainty associated with Cs_{sea} is the most important contributor to the uncertainty in the Cs-concentrations in water, in spite of the fact that the CV-value has been more than halved. The uncertainties associated with the tributary concentration are also more or less irrelevant in these tests. However, in river-mouth areas, they would be of greater significance.

• The uncertainties associated with the biouptake is also in this case the dominating uncertainty for the Cs-concentrations in fish. Those uncertainties can only be lowered if adequate model calibrations are performed.

2.5. The Simplified Model

These sensitivity and uncertainty analyses have revealed that the target predictions concerning Cs-concentrations in water and fish depend largely on two factors, i.e., the Cs-concentration in the sea outside the given coastal area and/or adjacent coastal areas and the ecological halflife of radiocesium in the sea. Uncertainties in these two driving variables dominate all other uncertainties, e.g., uncertainties regarding the surface water retention time, the settling velocity of particulate radiocesium, the distribution coefficient regulating the fluxes in dissolved and particulate phases, the catchment area influences, the sub-model for feeding habits, and the factors regulating biouptake and excretion of radiocesium in fish. This means that the conditions in the sea are of paramount importance for the conditions in the coastal area, even for enclosed coastal areas deep in archipelago areas. This also means that "one cannot save the coastal areas unless one first saves the sea". The coastal areas are often regarded as "nurseries and pantries" for the sea, but the conditions in the sea outside a given coastal area would regulate the transport of water pollutants to and from a given coastal area, and in most estuaries.

These results mean that one can produce a simplified version of the coastal model for radiocesium in which the relatively unimportant processes, factors and sub-models are omitted and/or simplified. The aim here is to derive such a simple model that can predict like the extensive model, but which is based on only the key processes, and the smallest possible number of driving variables.

The catchment-area model has been omitted. This means that catchment area, latitude, mean annual precipitation, outflow areas and soil permeability factor have been omitted from the panel of driving variables. This may imply that the simplified model underestimates the Cs-concentrations in water and biota in river-mouth areas. To compensate for this possible effect, the following "rules" have been included in the model concerning the ecological halflife for radiocesium in the sea. When the catchment-area sub-model is omitted, the secondary load from the catchment is neglected and this is compensated for by increasing the T_{sea}-value by 50%. This is done in the following simple manner.

Rule 1:

If the salinity in the coastal area ≤ 50% of the salinity in the sea, set T_{sea} to 1.5 multiplied by the default T_{sea}.

This also means that one must have some simple rules for setting the T_{sea}-value. This is done accordingly.

Rule 2:

For a Baltic coastal area, use T_{sea} = 3 years.

For a Mediterranean coastal area, use T_{sea} = 2 years.

For an Atlantic coastal area, use T_{sea} = 1 year.

The rationale for this is not very solid! The Baltic is, however, an enclosed sea, hence the high T_{sea}-value of 3 years (see Lozán et al., 1996). The Mediterranean is much bigger, but also enclosed, hence T_{sea} = 2. The currents along the Atlantic coast transport radiocesium from the tributaries along the coast. This means that the dilution is not as efficient as might be expected, hence T_{sea} = 1 year. This may not be true for many coastal areas, like the Irish west coast, where T_{sea} might be much shorter.

Since there are different water-temperature regimes in Baltic, Atlantic and Mediterranean coastal areas, the following default assumptions concerning surface and deep water temperatures have been included in the simplified model.

Rule 3:

For Baltic coastal areas, the surface water temperatures (°C) are given by:

Month:	1	2	3	4	5	6	7	8	9	10	11	12
MMST:	0	0	3	4	7	11	16	20	15	7	4	1

For Mediterranean coastal areas, the surface water temperatures (°C) are given by:

Month:	1	2	3	4	5	6	7	8	9	10	11	12
MMST:	6	6	6	6	7	11	16	20	15	7	6	6

For Atlantic coastal areas, the surface water temperatures (°C) are given by:

Month:	1	2	3	4	5	6	7	8	9	10	11	12
MMST:	9	9	9	11	13	15	18	22	17	14	12	10

The deep water temperatures cannot be handled without considering the thermal regime, i.e., the stratification of the water mass. The following rules have been used.

Rule 4:

If the coastal area is dimictic/monomictic (i.e., if it circulates twice or once a year), then $Td = 10 \cdot Ts$.

If the coastal area is polymictic (i.e., if it circulates often), then $Td = Ts$.

If the coastal area is meromictic (i.e., if it circulates very rarely, or never), then $Td = 100 \cdot Ts$.

From these presuppositions, the deep water temperatures may be estimated in the following manner.

Panel of model variables

Obliogatory driving variables

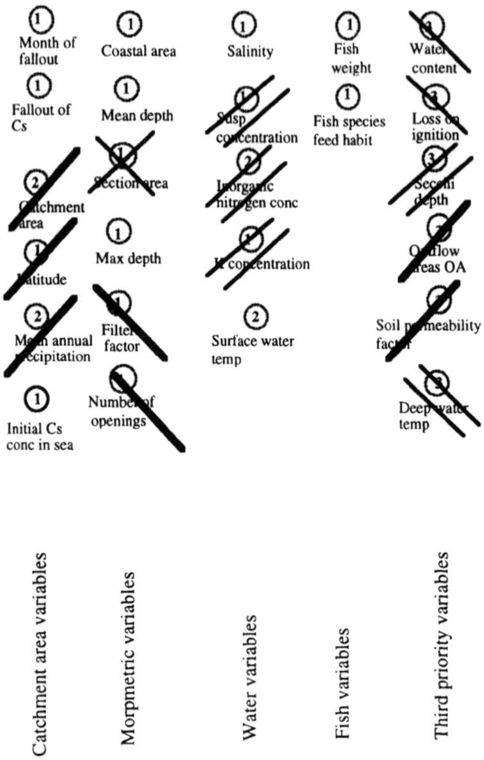

Guideline recommendations

= Driving variables omitted with the Q-model

Estuaries and coast types:

Rule 1:
If the salinity in the coastal area ≤ 50% of the salinity in the sea, set Tsea to 1.5 time default Tsea

Rule 2:
If Baltic coastal area, use Tsea = 3 years
If Mediterrainian coastal area, useTsea = 2 years.
If Atlantic coastal area, use Tsea = 1 year.

Rule 3:
If Baltic coastal area, surface water temperature given by:

Month:	1	2	3	4	5	6	7	8	9	10	11	12
Temp:	0	0	3	4	7	11	16	20	15	7	4	1

If Medeterrainian coastal area, surface water temperature given by:

Month:	1	2	3	4	5	6	7	8	9	10	11	12
Temp:	6	6	6	6	7	11	16	20	15	7	6	6

If Atlantic coastal area, surface water temperature given by:

Month:	1	2	3	4	5	6	7	8	9	10	11	12
Temp:	9	9	9	11	13	15	18	22	17	14	12	10

= Driving variables omitted with the Td-model

Rule 4, stratifications:
If the coastal area is dimictic/monomictic, then Td = 10*Ts.
If the coastal area is polymictic, then Td = Ts.
If the coastal area is meromictic, then Td = 100*Ts.

Rule 5, stratifications:
If the coastal area is dimictic/monomictic, then deep water temp given by:

Month:	1	2	3	4	5	6	7	8	9	10	11	12
Temp:	4	4	4	4	4.5	6	8	10	8	7	4	4

If the coastal area is polymictic, then deep water temp given surface water temperature.

If the coastal area is meromictic, then deep water temp given by:

Month:	1	2	3	4	5	6	7	8	9	10	11	12
Temp:	4	4	4	4	4.5	5	6	7	6	5	4	4

Rule 6, coastal openness:
If the coast is very enclosed, then At = 0.001*Area.
If the coast is semi-enclosed, then At = 0.01*Area.
If the coast line is open, then At = 0.1*Area.

Rule 7, coastal "energy filter"
If the coast is very enclosed, then Ff = 0.1.
If the coast is semi-enclosed, then Ff = 2
If the coast line is open, then Ff = 10.

= Driving variable omitted with the GS-model

Rule 8, trophic level:
Use table below to set (1) Secchi depth, (2) inorganic N-concentration, (3) K-concentration and (4) suspended particulate matter concentration.

Trophic level	Secchi (m)	Inorg-N (mg/m3)	K-conc (mg/l)	Susp (mg/l)
Oligot.	7	12	0.5	2
Mesot.	5	20	1	3
Eut.	2.5	35	2	5
Hypert.	1.5	50	3	7

= Driving variable omitted with the sediment model

Rule 9, sediment character:
Set water content of active sediment to 70% ww
Set loss on ignition of actuive A-sediments to 7% dw.

Rule 10, ET-areas:
ET = 10^(0.09+0.02*Dmax-0.24*Drel-2.75*√Vol).

Fig. 2.66. A summary of recommendations for the simplified coastal model and a comparison between the panel of driving variables for this model and the extensive model.

Rule 5:
If the coastal area is dimictic/monomictic, the mean monthly deep water temperatures (MMDT, °C) are given by:

Month:	1	2	3	4	5	6	7	8	9	10	11	12
MMDT:	4	4	4	4	4.5	6	8	10	8	7	4	4

If the coastal area is polymictic, the deep water temperatures (°C) are given by the surface water temperatures.

If the coastal area is meromictic, the deep water temperatures (°C) are given by:

Month:	1	2	3	4	5	6	7	8	9	10	11	12
MMDT:	4	4	4	4	4.5	5	6	7	6	5	4	4

This means that the sub-model for the deep water retention times is omitted. The following driving two variables can then be excluded: the filter factor and the number of openings (= section areas). It also means that the deep water temperatures can be eliminated from the panel of driving variables, since the values are given in a standardized way in the simplified model.

The fact that the filter factor is omitted means that one must also consider some rules for coastal openness and coastal "energy filters".

Rule 6:
If the coast is very enclosed, then set $At = 0.001 \cdot Area$.
If the coast is semi-enclosed, then $At = 0.01 \cdot Area$.
If the coast is open, then $At = 0.1 \cdot Area$.

This means that the section area (At) can also be omitted from the panel of driving variables. The values used for the filter factor are determined accordingly.

Rule 7:
If the coast is very enclosed, then $Ff = 0.1$.
If the coast is semi-enclosed, then $Ff = 2$
If the coast line is open, then $Ff = 10$.

This means that the user must decide if a coastal area is enclosed, semi-enclosed or open. By looking at the bathymetric map, the user must decide on the relationship between the section area (At) and the coastal area. If, e.g., the section area is ≥ 10% of the coastal area, the coast is open and the filter factor $Ff = 10$ is used in the model calculations.

The water variables needed to calculate gross sedimentation, i.e., inorganic nitrogen concentration (IN) and Secchi depth, have also been omitted from the simple model, along with suspended particulate matter and the K-concentration needed to calculate the partition coefficient and the settling velocity of particulate radiocesium. This has been achieved by using the following rules.

Rule 8:

Use the table below to set (1) Secchi depth, (2) inorganic N-concentration, (3) K-concentration and (4) suspended particulate matter concentration.

Trophic level	Secchi (m)	Inorg-N (mg/m^3)	K-conc (mg/l)	Susp (mg/l)
Oligot.	7	12	0.5	2
Mesot.	5	20	1	3
Eut.	2.5	35	2	5
Hypert.	1.5	50	3	7

If the user decides that the coastal area is oligotrophic, the model selects Secchi depth = 7 m, IN-conc. = 12 mg/m^3, K-conc. = 0.5 mg/l and SPM = 2 mg/l.

The next set of rules concerns the sediment characteristics.

Rule 9:

The water content of the active A-sediment is set to 70% ww for all coastal areas.

The loss on ignition (≈ organic content) of the active A-sediment is set to 7% dw for all coastal areas.

This means that in the simplified version the same values are used for all coastal areas for the water content and the organic content. Then these variables are not area-specific but model constants.

Rule 10:

A new simple regression model has been derived to estimate ET-areas from the few morphometric parameters accepted in the simple model for other reasons, i.e., mean depth, maximum depth and coastal area. This model gave an r^2-value of 0.54, as compared to 0.65 for the model presented in Table 2.4, which includes the mean filter factor. This model is based on the same data-set for 38 Baltic coastal areas. In the simplified model there is also an "If_then_else" statement which never allows ET to be smaller than 0.15 or larger than 1, the theoretical limits for ET. In the simple model ET is given by:

$$ET = 10^{(0.09+0.02 \cdot Dmax - 0.24 \cdot Drel - 2.75 \cdot \sqrt{Vol})} \tag{2.27}$$

Note that the volume is given in km^3, the mean depth in m and the relative depth as defined before.

Figure 2.66 gives a summay of all these simplifications.

The simple model is based on just the following driving variables: (1) fallout (Bq/month), (2) month of fallout, (3) size of coastal area (m^2), (4) mean depth (m), (5) maximum depth (m) and (6) salinity (‰). The user of the model must also decide which fish species to be predicted and the weight of the fish (see Table 1.8).

From these simple rules and the few obligatory driving variables almost the same predictions are obtain as by using the more comprehensive model. This is shown for the default coastal area in Figure 2.67. If the simple model is calibrated and validated, it has

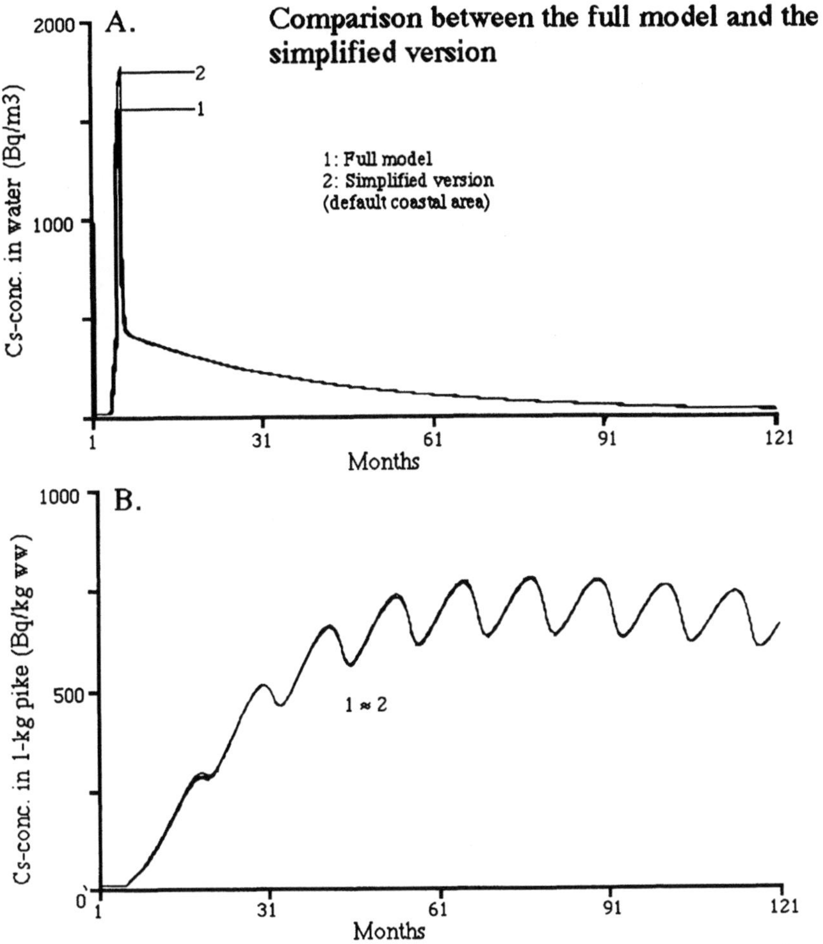

Fig. 2.67. A comparison between the extensive and the simplified model for the default coastal area for values for total Cs-concentrations in water (A) and Cs-concentrations in 1000 g ww piscivores (B).

the potential to become a useful tool in radioecology for predicting Cs-concentrations in water and fish shortly after a fallout.

CHAPTER 3

EPILOGUE

The ultimate aim of the work presented in this book is to produce meaningful modelling tools for practical water management. It is often stated that the quality of science is related to the possibilities for making meaningful predictions. Because of the complex nature of ecosystems, it has long been argued that it may be impossible to predict important target variables well, especially with more comprehensive dynamic models. Section 1 has demonstrated that such arguments are no longer valid. The key lies in the structuring and equations used for the predictive models.

There has been fantastic progress over the last decade of predictive ecosystem modelling. Ten years ago many models predicted target variables such as concentrations of toxins in fish with an uncertainty factor of about 10. Many people unfamiliar with ecosystem modelling probably still believe that the present status of ecosystem modelling is like it was ten years ago, and many traditional ecosystem modellers continue to believe that good, acceptable models yield uncertainty factors of around 2. The accident in Chernobyl has, in fact, provided science with an intriguing opportunity to study how the pulse of radiocesium is transported through ecosystem pathways, thus revealing the basic structure of these ecosystems, i.e., which are the vital and the less important pathways. It is paradoxical to conclude that the Chernobyl accident is, perhaps, the most important factor behind the revolution in predictive ecosystem modelling which lies behind the decrease in the uncertainty factor from 10 to 0.25-0.5.

It is generally more difficult to predict radiocesium concentrations in biota than in abiotic compartments (e.g., water and sediments). The new lake model for radiocesium discussed in section 1 has been validated against an independent set of data covering a wide range of lake characteristics. The validation for the Cs-concentrations in lake water gave an r^2-value of 0.92. The slope was 1.09. This regression was based on 61 monthly values governing a range from 0.001 to 4.5 Bq/l. This is a very good result for a validation. It actually means that the model predicts as good as the actual measurements - under the given conditions and in the given domain of the model. Naturally, when using models at the ecosystem level (for entire lakes, coastal areas, etc.), it is not possible to describe phenomena at the individual, organ or cellular levels.

The corresponding results for Cs-concentrations in lake fish are even better. The data-set comprises 185 monthly data for seven species of fish covering a very wide range, from 2 to over 30,000 Bq/kg ww. The r^2-value is 0.98 (for the log-transformed values). The slope is almost perfect. This demonstrates that the structuring of this model is very good. If such good predictions can be obtained for radiocesium for fish in lakes, then this holds great promise for other types of contaminants in future dynamic models. It also means that models of this type in the future are likely to become widely used and excellent tools in remedial contexts.

It would be tempting to conclude that this is the ultimate model for radiocesium in lakes, but such a conclusion would be premature. There are several empirical components in this model and there are always uncertainties associated with such empirical approaches, e.g., the soil permeability factor, the biouptake delay factor and the biomagnification factor. It would be a step forward, and it would add generic strength to this model, if it were possible to include more mechanistically based approaches for these empirical model components. The sub-model for the K-moderator is an example of such a mechanistically based approach. This approach is derived from the Michaelis-

Menten equation and the Nernst equation, and it provides the type of structure for this model that can help to explain its good predictive power.

The model for radiocesium in coastal areas is based on the same model structuring as the lake model. However, because of a lack of reliable empirical data from many coastal areas covering a wide range of coastal characteristics, this model has not been duly calibrated and validated. Section 2 presents the model as well as critical model tests (sensitivity and uncertainty analyses). It also gives brief backgrounds to the hydrodynamical and sedimentological processes regulating transport and biouptake of radiocesium (and most other pollutants) in coastal areas.

Several new approaches and sub-models have been presented for coastal modelling of water pollutants. These new features have been developed during the "evolution" of the overall model for radiocesium in coastal areas.

1. The method for defining the coastal area and setting the borderlines for the area by means of the "topographical bottle-neck" approach (where the exposure attains minimal values) has not, to the best of the author's knowledge, been used before in models of this type.

2. This definition of the coastal area is a necessary prerequisite for the determination of three fundamental process variables regulating the transport of radiocesium (and other pollutants as well): (1) The theoretical surface water retention time (Ts); (2) the theoretical deep water retention time (Td); and (3) the bottom dynamic conditions, or rather the area of the coast where advective transport processes regulate the internal loading of pollutants (i.e., resuspension from erosion- and transportation areas, ET), and areas with continuous sedimentation where diffusion dominates the internal loading (i.e., accumulation areas, A = 1-ET). This means that Ts, Td and ET can be determined very easily from standard bathymetric charts using GIS-methods. The approaches presented here for Ts, Td and ET have not been used before for dynamic coastal models of water pollutants. They are generic in the sense that they could be used for all types of pollutants and not just for radiocesium.

3. A new empirical sub-model is presented for gross sedimentation, regulating sediment age and the retention time of radiocesium in the active sediment layer, and hence the duration of the contamination. This empirical model is intended to replace an extensive dynamic model for gross sedimentation, since the model predictions for the target variables, Cs-concentrations in water and fish, are not generally so sensitive to the value for gross sedimentation. An important part of the model development has been to identify and rank the importance of the various sub-models and processes in the overall model and make justifiable simplifications whenever possible to keep the model small and practically useful. "Everything should be as simple as possible, but not simpler", according to Albert Einstein, and this model is constructed according to that philosophy.

4. Several new algorithms for such simplifications have been developed, many of which use the "dimensionless moderator technique", e.g., for the influence of suspended particulate matter and salinity (flocculation) on the settling velocity of particulate radiocesium and potassium concentration (based on the Michaelis-Menten and the Nernst equations) and salinity for the biouptake of radiocesium by fish.

5. A new simple regression equation for predicting the mean slope of coastal areas from information on the coastal area, mean depth and maximum depth has been presented. The mean slope is an important variable in the model for the theoretical deep water retention time.

Extensive sensitivity and uncertainty analyses have revealed that the target predictions concerning the Cs-concentrations in water and fish depend largely on two factors, i.e., the Cs-concentration in the sea outside the given coastal area and/or adjacent coastal areas and the ecological halflife of radiocesium in the sea. Uncertainties in these two variables dog all other uncertainties, e.g., regarding the surface and deep water retention times, the settling velocity of particulate radiocesium, the distribution coefficient regulating the fluxes in dissolved and particulate phases, the catchment-area influences, the sub-model for feeding habits, the bottom dynamic conditions, the sediment characteristics, and the factors regulating biouptake and excretion of radiocesium in fish. In short, the target predictions depend on these two factors which characterise the conditions outside the given coastal areas. This means that the conditions in the sea are of paramount importance for the conditions in the coastal area, even for enclosed coastal areas deep in archipelago areas. This also means that "one cannot save coastal areas unless one first saves the sea". The coastal areas are often regarded as "nurseries and pantries" for the sea, but the conditions in the sea outside a given coastal area regulate the transport of water pollutants to and from coastal areas. The coastal area, the outside sea and adjacent coastal areas should be regarded as a series of interrelated, communicating basins.

A simplified version of the radiocesium model has been constructed on the basis of results from the sensitivity and uncertainty analyses. All sub-models which contribute little to the predictive power of the model have been omitted and/or simplified in this model, which is based on few driving variables, such as fallout (Bq/month), month of fallout, size of coastal area (m^2), mean depth (m), maximum depth (m) and salinity (‰). The user of the model must also decide which fish species are to be predicted and the weight of the fish.

In the simplified model, there is a crude rule system for Baltic, Atlantic and Mediterranean coasts (which is used to set water temperatures). These simple rules are for stratification (the user must decide if the coastal waters are stratified or not, which is used to define mixing and volumes of surface water and deep water), for coastal openness (the user must decide if the coastal area is open, semi-enclosed or enclosed, as this regulates the inflow and outflow of water from the sea), and for the trophic level (the user must decide if the coastal area is oligo-, meso-, eu- and/or hypertrophic, which is used to calculate sedimentation, concentration of suspended particulate matter, inorganic nitrogen concentration, Secchi depth and potassium concentration of coastal waters). From these simple rules and the few obligatory driving variables almost the same predictions are obtain as can be achieved by using the more comprehensive model. The simple model has the potential to become a useful tool for predicting Cs-concentrations in water and fish very soon after a fallout. The extensive model may be regarded as a scientific tool for posing and testing working hypotheses on the relative roles of different processes in different coastal areas. Such information may be essential for getting realistic expectations of various remedial measures, such as coastal water fertilisation and dredging discussed in this book.

CHAPTER 4

LITERATURE REFERENCES

Abrahamsson, O. and Håkanson, L., 1997. Presentation and analysis of a model simulating the response of a potash treatment of a lake. J. Env. Radioactivity, 37:287-306.

Abrahamsson, O. and Håkanson, L., 1998. Modelling seasonal flow variability of European rivers. Ecol. Modelling, 114:49-58.

Alberts, J.J., Tilly, L.J. and Vigerstad, T.J., 1979. Seasonal cycling of cesium-137 in a reservoir. Science, 203:649-651.

Ambio, 1990. Special issue. Marine Eutrophication, 19:102-176.

Andersson, E., 1989. Incorporation of Cs-137 into fishes and other organisms. In: Feldt, W. (ed.). The radioecology of natural and artificial radionuclides. Verlag TUV, Köln. pp. 312-317.

Andersson, T., Forsgren, G., Håkanson, L., Malmgren, L. and Nilsson, Å., 1990. Radioactive caesium in fish in Swedish lakes after Chernobyl (in Swedish). SSI Report 90-04, 41 p.

Andersson, T., Håkanson, L., Kvarnäs, H. and Nilsson, Å., 1991. Measures against high levels of radiocesium in lake fish. Final report from the Liming-Mercury-Cesium project (in Swedish, Åtgärder mot höga halter av radioaktivt cesium i insjöfisk. Slutrapport - för cesiumdelen av projektet Kalkning-kvicksilver-cesium). Statens strålskyddsinst., SSI Report 91-07, Stockholm, 114 p.

Applegren, A., Bergström, U., Brittain, J., Gallego Diaz, E., Håkanson, L., Heling, R. and Monte, L., 1996. An outline of a model-based expert system to identify optimal remedial strategies for restoring contaminated aquatic ecosystems: The project "MOIRA". ENEA, ISSN/1120-5555, Roma, 45 p.

Beck, M.B. and Van Straten, G., 1983 (eds). Uncertainty, system identification and the prediction of water quality. Springer, Heidelberg, 387 p.

BIOMOVS (1990). Scenario A1. Mercury in aquatic ecosystems. Tech. Report 7. National Inst. of Radiation Prot., Stockholm, Sweden.

Black, V.S., 1957. Excretion and osmoregulation. In: The Physiology of Fishes, Brown, M.E. (ed.). Academic Press, New York, Vol. 1, pp. 163-205.

Bloesch, J. and Burns, N.M.,1980. A critical review of sedimentation trap technique. Schweiz. Z. Hydrol., 42:15-55.

Bloesch, J. and Uehlinger, U.,1986. Horizontal sedimentation differences in a eutrophic Swiss lake. Limnol. Oceanogr., 31:1094-1109.

Boers, P.C.M., Cappenberg, Th.E. and van Raaphorst, W., 1993 (eds). Proceedings of the Third International Workshop on Phosphorus in Sediments. Hydrobiologia, Vol. 253, 376 p.

Brittain, J. E., 1998. European fish species of radioecological importance. ENEA, Rome, RT/AMB/98/4, pp. 19-21.

Brittain, J., Håkanson, L., Bergström, U. and Bjørnstad, H. E., 1994. The significance of hydrological and catchment processes for the transport and biological uptake of radionuclides in northern aquatic ecosystems. Proceedings of the 10th International Northern Research Basins Symposium and Workshop, Spitsbergen, Norway.

Broberg, A. and Andersson, E., 1989. Circulation of caesium in limnic ecosystems (in Swedish). Inst. of Limnology, Uppsala univ., 30 p.

Burban, P.-Y., Lick, W. and Lick, J., 1989. The flocculation of fine-grained sediments in estuarine waters. J. Geophy. Res., 94:8223-8330.

Burban, P.-Y., Xu, Y.-J., McNeiel, J. and Lick, W. , 1990. Settling speeds of flocs in fresh water and seawater. J. Geophy. Res., 95:18,213-18,220.

Carlsson, L., Persson, J. and Håkanson, L., 1999. A management model to predict seasonal variability in oxygen concentration in thermally stratified coastal waters. Ecol. Modelling, 119: 117-134.

Carlsson, S., 1978. A model for the turnover of Cs-137 and potassium in pike (Esox Lucius). Health Phys., 35:549-554.

Chapra, S.C., 1980. Application of the phosphorus loading concept to the Great Lakes. In: Loehr, C., Martin, C.S. and Rast, W. (eds.), Phosphorus management strategies for lakes. Ann Arbor Science Publishers, Ann Arbor, pp. 135-152.

Chow, V. T., 1988. Applied hydrology. McGraw-Hill, Inc., 572 p

Comans, R.N.J. (ed.), 1998. Modelling fluxes and bioavailability of radiocesium and radiostrontium in freshwaters. Progress report from the ECOPRAQ project. Netherlands Energy Research Foundation.

Comans, R.N.J. and Hockley, D.E., 1992. Kinetics of cesium sorption on illite. Geochim. Cosmochim. Acta, 56:1157-1164.

Cox, D.C. and Baybutt, P., 1981. Methods for uncertainty analysis: a comparative survey. Risk Analysis, 1: 251-258.

Dahlgaard, H. (ed.), 1994. Nordic radioecology, the transfer of radionuclides through nordic ecosystems to man. Elsevier Science, Amsterdam. 483 p.

Desmet, G. et al. (eds), 1997. Freshwater and Estuarine Radioecology, Elsevier, Amsterdam, 501 p.

Dyer, J.L., 1972. Estuarine hydrography and sedimentation. Cambridge Univ. Press, Cambridge, 230 p.

Eberly, W.R., 1964. Further studies on the metalimnic oxygen maximum, with special reference to its occurrence throughout the world. Investigations of Indiana Lake Streams, Vol. 6.

Erel, Y. and Stolper, E.M., 1993. Modeling of rare-earth element partitioning between particles and solution in aquatic environments. Geochimica Cosmochimica Acta, 57:513-518.

Eriksson, E., 1974. Water, the carrier of chemicals (in Swedish). Forskning och Framsteg, 5:41-45.

Evans, R.D. and Håkanson, L., 1992. Measurement and prediction of sedimentation in small Swedish lakes. Hydrobiologia, 75:143-152.

Fernandez, J.A., Heredia, M.A., Garcia-Sanchez, M.J., Corisco, J.A.G., Vaz Carreiro, M.C., Diez de los Rios, A., 1997. Mechanisms of radiocesium uptake and accumulation in Riccia fluitans. In: Desmet, G. et al. (eds), Freshwater and Estuarine Radioecology, Elsevier, Amsterdam, pp. 329-338.

Fleishman, D.G., 1963. Accumulation of artificial radionuclides in freshwater fish. In: Radioecology, Klechkovskii, V.M., Polikarpov, G.G. and Aleksakhin, R.M. (editors). Wiley, New York, pp. 347-370.

Floderus, S., 1989. The effect of sediment resuspension on nitrogen cycling in the Kattegatt - variability in organic matter transport. Thesis, Uppsala Univ., UNGI Report, 71.

FRP. 1978. The sea; natural conditions and use (in Swedish, Havet; naturförhållanden och utnyttjande). Fysisk riksplanering (FRP), Bostadsdepartementet, Nr 7, 303 p.

Gilbert, R.O., 1987. Statistical methods for environmental pollution monitoring. Van Nostrand Reinold Co., New York.

Grip, H. and Rodhe, A., 1985. Vattnets väg från regn till bäck (The flow of water from rain to river; in Swedish). Forskningsrådens Förlagstjänst, Karlshamn, 156 p.

Gustafsson, Ö. and Gschwend, P. M., 1997. Aquatic colloids: Concepts, definitions and current challenges. Limnol. Oceanogr., 42:519-528.

Håkanson, L., 1977. The influence of wind, fetch, and water depth on the distribution of sediments in Lake Vänern, Sweden. Can. J. Earth Sci., 14:397-412.

Håkanson, L., 1981. A manual of lake morphometry. Springer, Heidelberg, 78 p.

Håkanson, L., 1986. The Swedish coastal zone project - sediment types and morphometry. In: Sly, P. G. (ed.), Sediment and water interactions, Springer-Verlag, New York, pp. 35-51.

Håkanson, L., 1990. A new functional view of the Baltic Sea. Ambio, Special Report 7:19.

Håkanson, L., 1991. Ecometric and Dynamic Modelling - exemplified by cesium in lakes after Chernobyl. Springer-Verlag, Berlin, 158 p.

Håkanson, L., 1995. Optimal size of predictive models. Ecol. Modelling, 78:195-204.

Håkanson, L., 1996. A new, simple, general technique to predict seasonal variability of river discharge and lake temperature for lake ecosystem models. Ecological Modelling 88:157-18.

Håkanson, L., 1997. Modelling of radiocesium in lakes - on predictive power and lessons for the future. In: Desmet, G. et al. (eds), Freshwater and Estuarine Radioecology, Elsevier, Amsterdam, pp. 3-45.

Håkanson, L., 1998a. A compilation of empirical data and variations in data concerning radiocesium in water, sediments and fish in European lakes after Chernobyl. J. Env. Radioactivity (in press).

Håkanson, L., 1999. Water Pollution - methods and criteria to rank, model and remediate chemical threats to aquatic ecosystems. Backhuys Publishers, Leiden, 277 p.

Håkanson, L. and Andersson, T., 1992. Remedial measures against radioactive caesium in Swedish lake fish after Chernobyl. Aquatic Sci., 54:141-164.

Håkanson, L. and Jansson, M., 1983. Principles of Lake Sedimentology. Springer, Berlin, 316 p.

Håkanson, L., Kulinski, I. and Kvarnäs, H., 1984. Water dynamics and bottom dynamics in the coastal zone (in Swedish, Vattendynamik och bottendynamik i kustzonen). National Swedish Environmental Protection Agency, SNV PM 1905, Solna, 228 p.

Håkanson, L., Kvarnäs, H. and Karlsson, B., 1986. Coastal morphometry as regulator of water exchange - a Swedish example. Estuarine, Coastal & Shelf Science, 23:1-15.

Håkanson, L. and Peters, R.H., 1995. Predictive Limnology. Methods for predictive modelling. SPB Academic Publishing, Amsterdam, 464 p.

Håkanson, L. and Rosenberg, R., 1985. Practical coastal ecology (in Swedish, Praktisk kustekologi). Swedish Environmental Protection Agency, PM 1987, Solna, 110 p.

Håkanson, L., Andersson, T., Neumann, G., Nilsson, Å. and Notter, M., 1988. Cesium in perch in lakes from northern Sweden after Chernobyl - present situation, causal relationships, the future (In Swedish, Cesium i abborre i norrländska sjöar efter Tjernobyl - läget, orsakssamband, framtiden). SNV Report 3497, 136 p.

Håkanson, L., Andersson, T. and Nilsson, Å., 1992. Radioactive caesium in fish in Swedish lakes 1986-1988 - general pattern related to fallout and lake characteristics. J. Env. Radioactivity, 15:207-229.

Håkanson, L., Jonsson, B., Jonsson, P. and Martinsen, K., 1988b. Impact areas of chlorinated organics from paper and pulp mill (in Swedish, Påverkansområden av klorerat organiskt material från massablekerier. Slutrapport). Swedish Environmental Protection Agency, Report 3522, Stockholm, 165 p.

Håkanson, L., Brittain, J., Monte, L., Heling, R., Bergström, U. and Suolanen, V., 1996a . Modelling of radiocesium in lakes - the VAMP-model. J. Environ. Radioactivity, 33:255-308.

Håkanson, L., Brittain, J., Monte, L., Heling, R., Bergström, U. and Suolanen, V., 1996b. Modelling of radiocesium in lakes - lake sensitivity and remedial strategies. J. Environ. Radioactivity, 33:1-25.

Håkanson, L., Abrahamsson, O., Ottosson, F., and Johansson, T., 1998a. Presentation and analysis of a model simulating the response of lake fertilisation. J. Environ. Radioactivity, 41:343-380.

Hamby, D.M., 1995. A comparison of sensitivity analysis techniques. Health Physics, 68:195-204.

Hilton, J., 1985. A conceptual framework for predicting the occurrence of sediment focusing and sediment redistribution in small lakes. Limnol. Oceanogr., 30:1131-1143.

Hilton, J., 1997. Aquatic radioecology post Chernobyl - a review of the past and a look to the future. In: Desmet, G. et al. (eds), Freshwater and Estuarine Radioecology, Elsevier, Amsterdam, pp. 47-73.

Hinton, T. G., 1993. Sensitivity analysis of ecosys-87: an emphasis on the ingestion pathway as a function of radionuclide and type of disposition. Health Physics, 66:513-531.

IAEA, 1988. International Atomic Energy Agency. Assessing the impact of deep sea disposal of low level radioactive waste on living marine resources. Tech. Rep. No. 288, Vienna.

IAEA, 1992. International Atomic Energy Agency. Effects of ionizing radiation on plants and animals at levels implied by current radiation protection standards. Tech. Rep. No. 332, Vienna.

IAEA, 1999. International Atomic Energy Agency. Modelling of Radiocesium in lakes. Tec. Doc. (in print), Vienna.

ICME, 1995. The International Council on Metals and the Environment. Persistence, bioaccumulation and toxicity of metals and metal compounds. Parametrix Inc., ISBN-1-895720-07-9.

ICRP, 1977. International Commission of Radiological Protection. Recommendations of the International Commission of Radiological Protection, Publication 26, Pergamon Press, Oxford.

Jansson, M., Hayman, U. and Forsberg, C., 1981. Acid lakes and "biological buffering" (in Swedish). Vatten, 37: 241-251.

Jimenez, F. and Gallego, E., 1998. Effects of radiation on aquatic organisms. Preprint, from the MOIRA project, Univ. de Politech. de Madrid, Spain, 9 p.

Jonsson, A., 1997. Whole lake metabolism of allochthonous organic material and the limiting nutrient concept in Lake Örträsket, a large humic lake in northern Sweden. Diss. Dept. of Physical Geography, Univ. Univ., Sweden, ISBN 91-7191-382-3.

Jonsson, P., 1992. Large-scale changes of contaminants in Baltic Sea sediments during the twentieth century. Thesis, Uppsala Univ., Sweden.

Jørgensen, S.E., Kamp-Nielsen, L. and Jörgensen, L.A., 1986. Examination of the generality of eutrophication models. Ecol. Modelling, 32:251-266.

Jørgensen, S.E. and Johnsen, J., 1989. Principles of environmental science and technology (2nd edition). Studies in environmental science, 33. Elsevier, Amsterdam, 628 p.

Konitzer, K. and Meili, M., 1997. Redistribution of sedimentary Cs-137 in small Swedish lakes after the Chernobyl fallout 1986. In: Desmet, G. (et al., editors), Freshwater and Estuarine Radioecology, pp. 167-172, Elsevier, Amsterdam.

Konoplev, A., Bulgakov, A., Hilton, J., Comans, R. and Popov, V., 1997. Long-term kinetics of radiocesium fixation by soils. In: Desmet, G. (et al., editors), Freshwater and Estuarine Radioecology, pp. 173-182, Elsevier, Amsterdam.

Kranck, K., 1973. Flocculation of suspended sediment in the sea. Nature, 246:348-350.

Kranck, K., 1979. Particle matter grain-size characteristics and flocculation in a partially mixed estuary. Sedimentology, 28:107-114.

Lick, W., Lick, J. and Ziegler, C.K., 1992. Flocculation and its effect on the vertical transport of fine-grained sediments. Hydrobiologia, 235/236:1-16.

Lindström, M., Håkanson, L., Abrahamsson, O. and Johansson, H., 1999. An empirical model for prediction of lake water suspended matter. Ecol. Modelling, 121:185-198.

Lozán, J.L., Lampe, R., Mattäus, W., Rachor, E., Rumohr. H. und von Westernhagen, H. (editors), 1996. Warnsignale aus der Ostsee. Pary, Buchverlag, Berlin, 385 p.

McCave, I.N., 1981. Location of coastal accumulations of fine sediments around the southern North Sea. Rapp. P. v. Reun. Int. Explor. Mer., 181:15-27.

Madruga, M.J. and Cremers, A., 1997. On the differential binding mechanisms of radiostrontium and radiocesium in sediments. In: Desmet, G. (et al., editors), Freshwater and Estuarine Radioecology, pp. 207-216, Elsevier, Amsterdam.

Moberg, L. (ed.), 1991. The Chernobyl Fallout in Sweden. The Swed. Rad. Prot. Inst., Stockholm, 633 p.

Monte, L., 1995. A simple formula to predict approximate initial contamination of lake water following a pulse deposition of radionuclide. Health Physics, Vol. 68, Nr. 3.

Monte, L., 1996. Collective models in environmental science. Sci. Tot. Environ., 192:41-47.

Monte, L., Håkanson, L and Brittain, J., 1997. Prototype models for the MOIRA computerised system. ENEA, ISSN/1120-5555, Roma, 90 p.

Muir Wood, A.M., 1969. Coastal hydraulics. Macmillan, London, 187 p.

Nixon, S.W., 1990. Marine eutrophication: a growing international problem. Ambio, 3:101.

Nyström, U., 1985. Transit time distribution of water in two small forested catchments. Ecol. Bull., Stockholm, 37:98-100.

Ottosson, F. and Håkanson, L., 1997. Presentation and analysis of a model simulating the pH response of a lake liming. Ecol. Modelling, 105:89-111.

Ottosson, F. and Abrahamsson, O., 1998. Presentation and analysis of a model simulating epilimnetic and hypolimnetic temperatures in lakes. Ecol. Modelling, 110:223-253.

Persson, C., Rodhe, H. and De Geer, L.-E., 1987. The Chernobyl accident - A meteorological analysis of how radionuclides reached and were deposited in Sweden. Ambio, 16:20-31.

Persson, J. and Håkanson, L., 1995. Prediction of bottom dynamic conditions in coastal waters. Mar. and Freshw. Res., 46:359-371.

Persson, J. and Håkanson, L., 1996. A simple empirical model to predict deepwater turnover time in coastal waters. Can. J. Fish. Aq. Sci., 53:1236-1245.

Persson, J., Håkanson, L. and Pilesjö, P., 1994. Prediction of surface water turnover time in coastal waters using digital bathymetric information. Environmentrics, 5:433-449.

Peters, R.H., 1986. The role of prediction in limnology. Limnol. Oceanogr., 31:1143-1159.

Peters, R.H., 1991. A Critique for Ecology. Cambridge Univ. Press, Cambridge, 366 p.

Pilesjö, P., Persson, J. and Håkanson, L., 1991. Digital bathymetric information for calculations of coast morphometrical parameters and surface water retention time (in Swedish with English summary). Swedish Environmental Protection Agency, Report 3916, 76 p.

Postma, H., 1982. Sediment transport and sedimentation. In: Olausson, E. and Cato, I. (eds), 1982, Chemistry and biogeochemistry of estuaries, Wiley & Sons, Chichester, pp. 153-186.

Prairie, Y., 1996. Evaluating the predictive power of regression models. Can. J. Fish. Aquat. Sci., 53:490-492.

Pustelnikov, O.S., 1977. Geochemical features of suspended matter in connection with recent sedimentation processes in the Baltic Sea. Ambio, 5:157-162.

Redfield, A.C., 1958. The biological control of chemical factors in the environment. Am. Sci., 46:205-222.

Remane, A., 1934. Die Brackwasserfauna. Verh. Dt. Zool. Ges., 36:34-74.

Riise, G., Björnstad, H.E., Oughton, D.H. and Salbu, 1990. A study on radionuclide associations with soil components using sequential extraction procedure. J. Radioanal. Nucl. Chem., 142:531-538.

Rodhe, A., 1987. The origin of streamwater traced by oxygen-18. Thesis, Uppsala univ., Series A, No. 41, 260 p.

Rose, K.A., McLean, R.I. and Summers, J.K., 1989. Development and Monte Carlo analysis of an oyster bioaccumulation model applied to biomonitoring. Ecol. Modelling, 45:111-132.

Rowan, D.J. and Rasmussen, J.B., 1994a. Bioaccumulation of radiocesium by fish: the influence of physicochemical factors and trophic structure. Can. J. Fish. Aquat. Sci., 51:2388-2410.

Rowan, D.J. and Rasmussen, J.B., 1995. The elimination of radiocesium from fish. J. Appl. Ecology., 32:739-744.

Ryther, J.H. and Dunstan, W.M., 1971. Nitrogen, phosphorus, and eutrophication in the coastal marine environment. Science, 171:1008-1013.

Santschi, P.H. and Honeyman, B.D., 1991. Radioisotopes as tracers for the interactions between trace elements, colloids and particles in natural waters. In: Vernet, J.-P., ed. Heavy Metals in the Environment, Elsevier Amsterdam.

Schindler, D.W., 1977. Evolution of phosphorus limitation in lakes. Science, 195:260-262.

Schindler, D.W., 1978. Factors regulating phytoplankton production and standing crop in the world's freshwaters. Limnol. Oceanogr., 23:478-486.

Schulze, E.-D. and Zwölfer, H. (eds), 1987. Potentials and Limitations of Ecosystem Analysis. Springer, Heidelberg, 435 p.

Seibold, E. and Berger, W.H., 1982. The sea floor. Springer-Verlag, Heidelberg, 288 p.

SMHI, 1995a. Vattenföring i Sverige. Del 1. Vattendrag till Bottenviken. [Water discharge in Sweden. Part 1. Water paths to the northern Gulf of Bothnia]. SMHI, Norrköping.

SMHI, 1995b. Vattenföring i Sverige. Del 2. Vattendrag till Bottenhavet. [Water discharge in Sweden. Part 2. Water paths to the Sea of Bothnia), SMHI, Norrköping.

SMHI, 1995c. Vattenföring i Sverige. Del 3. Vattendrag till Egentliga Östersjön. [Water discharge in Sweden. Part 4. Water paths to the Baltic Proper]. SMHI, Norrköping.

SMHI, 1995d. Vattenföring i Sverige. Del 4. Vattendrag till Västerhavet. [Water discharge in Sweden. Part 44. Water paths to the Western sea]. SMHI, Norrköping.

Smith, J.T., Leonard, D.R.P., Hilton, J. and Appleby, P.G., 1997. Towards a generalized model for the primary and secondary contamination of lakes by Chernobyl-driven radiocesium. Health Physics, 72: 880-892.

Stanley, D.J. and Swift, D.J., (eds), 1976. Marine sediment transport and environmental management. Wiley & Sons, New York, 602 p.

Straskraba, M. and Gnauck, A., 1985. Freshwater Ecosystems. Modelling and Simulation. Developments in Environmental Modelling, 8. Elsevier, Amsterdam, 310 p.

Tiwari, J.L. and Hobbie, J.E., 1976. Random differential equations as models of ecosystems. Monte Carlo simulation approach. Math. Biosci., 28:25-44.

UNESCO, 1993. Discharge of selected rivers of the world. Volume II (part II). Monthly and annual discharges recorded at various selected stations 20-years catalogue (1965-1984). Gidrometeoizdat, St. Petersburg.

USNRC, 1978. United States Nuclear Regulatory Commission. Liquid Pathway Generic Study. Impacts of accidental radioactive releases to the hydrosphere from floating and land-based nuclear power plants. NUREG-0440, USNRC, Washington.Vemuri, V., 1978. Modeling of complex systems. Academic Press, New York, 448 p.

Voipio, A. (ed.), 1981. The Baltic Sea. Elsevier Oceanographic Series, Amsterdam, 418 p.

Vollenweider, R.A., 1968. The scientific basis of lake eutrophication, with particular reference to phosphorus and nitrogen as eutrophication factors. Tech. Rep. DAS/DSI/68.27, OECD, Paris, 159 pp.

Wallin, M., Håkanson, L. and Persson, J., 1992. Load models for nutrients in coastal areas, especially from fish farms (in Swedish with English summary). Nordiska ministerrådet, 1992:502, Copenhagen, 207 p.

Wetzel, R.G., 1983. Limnology. Saunders College Publ., 767 p.

Whicker, F.W., 1997. Measurement quantities and units. Preprint from an International Workshop on Measuring Radionuclides in the Environment: Radiological Quantities and Sampling Designs, Bad Honnef, Germany, Nov., 1997.

Whicker, F.W. and Schultz, V., 1982. Radioecology: Nuclear Energy and the Environment. Volume 1, CRC Press, Boca Raton, 228 p.

Worley, B.A., 1987. Deterministic uncertainty analysis. Oak Ridge National Laboratory Report ORNL-6428, Oak Ridge, U.S.A., 53 p.

CHAPTER 5

APPENDIX

5.1. Equations for the Lake Model

A compilation of all equations in the radiocesium model for lakes using data for Lake Siggefora, Sweden.

Compartment: A-areas.

Amount_in_A_areas(t) = Amount_in_A_areas(t - dt) + (Sed_on_A + Advection_to_A - Diffusion - To_deep_sediments) * dt
INFLOWS:
Sed_on_A = Cs_amount_in_lake_water*Sedimentation_rate_of_part_Cs*(1-ET_areas)
Advection_to_A = (Amount_in_ET_areas*Form_factor/3)/Age_of_ET_sediments
OUTFLOWS:
Diffusion = if Amount_in_A_areas > 0 then Amount_in_A_areas*Diffusion_rate else 0
To_deep_sediments = if Amount_in_A_areas < 0 then 0 else Amount_in_A_areas*(1/Age_of_surface_sediments+Physical_decay_rate_for_Cs).

Compartment: ET-areas.

Amount_in_ET_areas(t) = Amount_in_ET_areas(t - dt) + (Sed_on_ET - Advection_to_wat - Advection_to_A - Decay_from_ET) * dt
INFLOWS:
Sed_on_ET = Cs_amount_in_lake_water*Sedimentation_rate_of_part_Cs*ET_areas
OUTFLOWS:
Advection_to_wat = ((Amount_in_ET_areas*(1-Form_factor/3))/Age_of_ET_sediments)
Advection_to_A = (Amount_in_ET_areas*Form_factor/3)/Age_of_ET_sediments
Decay_from_ET = Amount_in_ET_areas*Physical_decay_rate_for_Cs

Compartment: inflow areas.

Amount_in_inflow_areas(t) = Amount_in_inflow_areas(t - dt) + (Deposition_on_IA - Secondary_input_from_IA - Decay_from_IA - From_IA_to_OA) * dt
INFLOWS:
Deposition_on_IA = Fallout_of_Cs*Catchment_area*(1-Outflow_areas_OA)
OUTFLOWS:
Secondary_input_from_IA = ((1-Distribution_coeff)*Amount_in_inflow_areas*Default_runoff_rate_for_OA)/(12*Soil_permeability_factor)
Decay_from_IA = Amount_in_inflow_areas*Physical_decay_rate_for_Cs
From_IA_to_OA = (Distribution_coeff*Amount_in_inflow_areas*Default_runoff_rate_for_OA)/(12*Soil_permeability_factor)

Compartment: outflow areas.

Amount_in_outflow_areas(t) = Amount_in_outflow_areas(t - dt) + (Deposition_on_OA + From_IA_to_OA - Secondary_input_from_OA - Decay_from_OA) * dt
INFLOWS:
Deposition_on_OA = Fallout_of_Cs*Catchment_area*Outflow_areas_OA
From_IA_to_OA = (Distribution_coeff*Amount_in_inflow_areas*Default_runoff_rate_for_OA)/(12*Soil_permeability_factor)
OUTFLOWS:
Secondary_input_from_OA = Amount_in_outflow_areas*Modified_runoff_rate
Decay_from_OA = Amount_in_outflow_areas*Physical_decay_rate_for_Cs

Compartment: lake water.

Cs_amount_in_lake_water(t) = Cs_amount_in_lake_water(t - dt) + (Direct_lake_fallout + Diffusion + Secondary_input_from_OA + Secondary_input_from_IA + Advection_to_wat - Lake_outflow - Sed_on_A - Sed_on_ET) * dt
INFLOWS:
Direct_lake_fallout = (Fallout_of_Cs*Lake_area)
Diffusion = if Amount_in_A_areas > 0 then Amount_in_A_areas*Diffusion_rate else 0
Secondary_input_from_OA = Amount_in_outflow_areas*Modified_runoff_rate
Secondary_input_from_IA = ((1-Distribution_coeff)*Amount_in_inflow_areas*Default_runoff_rate_for_OA)/(12*Soil_permeability_factor)
Advection_to_wat = ((Amount_in_ET_areas*(1-Form_factor/3))/Age_of_ET_sediments)
OUTFLOWS:
Lake_outflow = (Physical_decay_rate_for_Cs+Retention_rate_in_lake_water)*Cs_amount_in_lake_water
Sed_on_A = Cs_amount_in_lake_water*Sedimentation_rate_of_part_Cs*(1-ET_areas)
Sed_on_ET = Cs_amount_in_lake_water*Sedimentation_rate_of_part_Cs*ET_areas

Compartment: fish.

Cs_concentration_in_fish(t) = Cs_concentration_in_fish(t - dt) + (Biouptake_fish - Excretion_fish) * dt
INFLOWS:
Biouptake_fish =
All_prod_moderator*Aut_prod_moderator*K_moderator*Biomagnification_factor*SMTH(Dissolved_conc+Particulate_phase,Biouptake_delay_factor,Dissolved_conc+Particulate_phase)
OUTFLOWS:
Excretion_fish = Cs_concentration_in_fish*(Biological_halflife+Physical_decay_rate_for_Cs)

Sub-models and driving variables:

Active_vol = if Epi_vol < 0.5*Lake_area then 0.5*Lake_area else Epi_vol
Age_of_ET_sediments = 0.5*2*12/1.386
Age_of_surface_sediments =
Bioturbation_factor*Depth_of_active_sed*(0.1*1.05*12)/(Gross_sedimentation*365*0.7*10^(-8))
All_prod_moderator = if Outflow_areas_OA < 0.1 then 1 else 0.1/Outflow_areas_OA
Altitude = (75+1)
Aut_prod_moderator = if TotalP > 14 then 8 else (22-TotalP)
Averaging_function = 500000/(Precipitation*(Altitude*Latitude*Catchment_area)^(1/4)))

Biological_halflife = if Fish_weight < 5 then 30*(EXP(-6.583-0.111*LOGN(Fish_weight*5)+0.093*Epi_temp+0.326*Steady_state)) else 1*30*(EXP(-6.583-0.111*LOGN(Fish_weight)+0.093*Epi_temp+0.326*Steady_state))
Biological_halflife_for_Cs137_in_plants = 1*0.55/30
Biomagnification_factor = if Fish_weight > 10 then (if Feed_habit = 0.005 then 250/Fish_weight else 0) + (if Feed_habit = 0.5 then 2 else 0) + (if Feed_habit = 0.05 then 3 else 0) + (if Feed_habit = 0.02 then 2.5 else 0)+ (if Feed_habit = 0.1 then 3.5 else 0) + (if Feed_habit = 0.125 then 4 else 0)+(if Feed_habit = 0.25 then 2.5 else 0)+ (if Feed_habit = 0.125/2 then 3.5 else 0) + (if Feed_habit = 0.01 then 1.5 else 0) else 25
Biomagnification_factor_for_plants = 2.5/10
Bioturbation_factor = if Gross_sedimentation > 2000 then 1 else (4-0.02*(Gross_sedimentation/100-1))
Biouptake_delay_factor = if Fish_weight > 100 then (Fish_weight/100) else 1
Bulk_density = 100*2.6/(100+(Water_content+Loss_on_ignition)*(2.6-1))

Catchment_area = 21*10^6
Conc_in_A_sed = Amount_in_A_areas/((10^6)*Volume_of_A_sed*Bulk_density*(1-Water_content/100))
Continentality = 100

Dcrit = if DTA < 0.5 then 0.5 else DTA
Default_fall_velocity_for_Cs = (12/12)*(1+0.75*(Susp_matter/50-1))
Default_runoff_rate_for_OA = 0.04*(Precipitation/650)^2
Depth_of_active_sed = 0.02
Diffusion_rate = if Bioturbation_factor = 1 then (20*0.002/12)*(Gross_sedimentation/100)*(Hypo_temp/4) else (0.002/12)*(Gross_sedimentation/100)*(Hypo_temp/4)
Dissolved_conc = Total_Cs_in_water*1/(1+Lake_Kd*Susp_matter/1000000)
Dissolved_fraction = 1/(1+Lake_Kd*Susp_matter/1000000)
Distribution_coeff = 0.5
Distribution coefficient for resuspension (DC)
DC2(t) = DC2(t - dt) + (DC1 - DC) * dt
INFLOWS:
DC1 = if (Advection_to_wat+Advection_to_A)/(Sed_on_ET+Sed_on_A) > 0.999 then 0.999 else (Advection_to_wat+Advection_to_A)/(Sed_on_ET+Sed_on_A)
OUTFLOWS:
DC = DC2
DTA = if (45.7*SQRT(Lake_area*10^(-6))/(21.4+SQRT(Lake_area*10^(-6)))) > Max_depth then Max_depth else (45.7*SQRT(Lake_area*10^(-6))/(21.4+SQRT(Lake_area*10^(-6))))
Dynamic_ratio = (Lake_area*10^(-6))^0.5/Mean_depth

E_1 = DELAY(MMET_theor2,0.5)
E_2 = DELAY(MMET_theor2,1)
Emp_mean_annual_Q = 0
Epi_area = Lake_area*0.01*(100*EXP(-3+Form_factor^1.5)*(Max_depth-Epi_depth)/(Epi_depth+Max_depth*EXP(-3+Form_factor^1.5)))^(0.5/Form_factor)*10^(2-1/Form_factor)
Epi_depth = 0.25*ABS(Epi_temp-Hypo_temp)
Epi_temp = MMET= if abs(MMET_theor2-MMHT_theor2)<=5 then
if (MMET_theor2+MMHT_theor2)<4.1 then 0
else MMET_theor2/2+MMHT_theor2/2
else MMET_theor2
Epi_vol = if Epi_depth = 0 then Mean_depth*Epi_area else (Lake_area-Epi_area)*Epi_depth/3 + Epi_area*Epi_depth

ET_areas = if ET_limit_4 > 1 then 1 else ET_limit_4
ET_areas_emp = 0.24
ET_limit_1 = 1-(Lake_area*0.01*(100*EXP(-3+Form_factor^1.5)*(Max_depth-Dcrit)/(Dcrit+Max_depth*EXP(-3+Form_factor^1.5)))^(0.5/Form_factor)*10^(2-1/Form_factor))/Lake_area
ET_limit_2 = if ET_limit_1 > 0.99 then 0.99 else ET_limit_1
ET_limit_3 = if ET_limit_2 < 0.15 then 0.15 else ET_limit_2
ET_limit_4 = if Lake_area > 10^6 then 0.25*Dynamic_ratio*41^(0.061/Dynamic_ratio) else ET_limit_3

Feed_habit = if Fish_weight > 10 then 1/8 else 0.005
Fish_weight = 1000
Form_factor = 3*Mean_depth/Max_depth

Gross_sedimentation = (Form_factor/3)*10^(-1.56-0.53*log10(Rel_depth)+0.4445*pH+0.8524*log10(TotalP))

H_1 = DELAY(MMHT_theor2,0.5)
H_2 = DELAY(MMHT_theor2,1)
HypoT = MMHT = if (abs(MMET_theor2-MMHT_theor2)>=5) or ((MMET_theor2+MMHT_theor2)/2<=4) then
(if E_1/2+H_1/2>=MMHT_theor2 then
(if abs(MMET_theor2-MMHT_theor2)<=8 then
(if ((MMET_theor2+MMHT_theor2)/2<=4) and
(MMET_theor2-E_1<0) then 4
else (E_1+H_1)/2)
else (E_2+H_2)/2)
else 4)
else (MMET_theor2+MMHT_theor2)/2

K_concentration = 0.6/0.0391
K_moderator = 300/(4.5+K_concentration^0.75)

Lake_area = 0.73*10^6 m^2
Lake_Kd = 800000/(K_concentration*0.0391)
Latitude = 60 °N
Loss_on_ignition = 1.3 % dw

MAET = if 44-(750/(90-Latitude^0.85))^1.29-0.1*Altitude^0.5-0.25*(Continentality^0.9+500)^0.52<4 then 4 else 44-(750/(90-Latitude^0.85))^1.29-0.1*Altitude^0.5-0.25*(Continentality^0.9+500)^0.52
Max_depth = 11 m
Mean_annual_Q = if Emp_mean_annual_Q=0 then 0.38*Precipitation*Catchment_area*10^-6/3600/24/365.25*1000 else Emp_mean_annual_Q
Mean_depth = 4.2 m
MMET_theor = MAET+smth1(Temp_norm,MAET/6*(Volume*10^(-6))^0.1,1)
MMET_theor2 = if MMET_theor < 0 then 0 else MMET_theor
MMHT_theor = smth1(MMET_theor,MAET^0.5,0.51*MAET)/(0.5/(1.1/(Mean_depth+0.1)+0.2))
MMHT_theor2 = if MMHT_theor < 4 then 4 else if (MAET<4) or (MAET > 17) or (Mean_depth < 2) then MAET else MMHT_theor
Modified_runoff_rate = Seasonal_moderator_for__Q*Default_runoff_rate_for_OA/(12*(sqrt(DELAY(TIME,Month_of_fallout,1))))
Month_of_fallout = 5
Monthly_water_discharge = Water_discharge*Seasonal_moderator_for__Q

Outflow_areas_OA = 0.1

Particulate_conc = 1000*(Total_Cs_in_water-Dissolved_conc)/Susp_matter

Particulate_phase = Feed_habit*((Total_Cs_in_water-Dissolved_conc)+0.001*(Amount_in_A_areas+Amount_in_ET_areas)/Volume)
pH = 7.2
Physical_decay_rate_for_Cs = 0.693/(30.2*12)
Precipitation = 750 mm/yr

Rel_depth = (Max_depth*(3.1416^0.5))/(20*(Lake_area*10^(-6))^0.5)
Retention_rate_in_lake_water = if Theoretical_water_ret_time < 1 then 1.386/Theoretical_water_ret_time else (if Lake_area < 0.2*10^6 then 1.386/Theoretical_water_ret_time^((50/(Theoretical_water_ret_time+50-1) + 0.5)/1.5)+ else 1.386/Theoretical_water_ret_time^((10/(Theoretical_water_ret_time+10-1) + 0.5)/1.5)))

Seasonal_moderator_for__Q = 1+0.526*((Latitude-35)^2.18/35^2.18*Seas_norm_Latmin+(1-(Latitude-35)^2.18/35^2.18)*Seas_norm_Latmax) +

0.421*(Altitude^0.51/1000^0.51*Seas_norm_Altmin+(1-Altitude^0.51/1000^0.51)*Seas_norm_Altmax) + 0.265*(Mean_annual_Q^0.22/5000^0.22*Seas_norm_Qmin + (1-Mean_annual_Q^0.22/5000^0.22)*Seas_norm_Qmax)
Sedimentation_rate_of_part_Cs = (((1-Dissolved_fraction)*(1-DC)*Default_fall_velocity_for_Cs)/Mean_depth)+(((1-Dissolved_fraction)*(DC)*2*Default_fall_velocity_for_Cs)/Mean_depth)
Soil_permeability_factor = 80
Steady_state = 1
Susp_matter = 2.1+0*10^(-2.0007+1.1803*LOG10(TotalP)+0.1335*pH+0.2804*LOG10(Dynamic_ratio)), mg/l

Theoretical_water_ret_time = 0.78*12 months
Total_Cs_in_water = 0.001*Cs_amount_in_lake_water/Volume
TotalP = 10 μg/l

Volume = Mean_depth*Lake_area m^3
Volume_of_A_sed = Lake_area*(1-ET_areas)*Depth_of_active_sed/2

Water_content = 95 % ww
Water_discharge = Volume/Theoretical_water_ret_time

Fallout_of_Cs = 30000 Bq/m^2 month 5.

Seas_norm_Altmax = GRAPH(MOD(time, 12))
(1.00, -0.97), (2.00, -0.98), (3.00, -0.58), (4.00, -0.69), (5.00, 2.11), (6.00, 1.87), (7.00, 0.51), (8.00, 0.07), (9.00, 0.03), (10.0, -0.06), (11.0, -0.62), (12.0, -0.68)
Seas_norm_Altmin = GRAPH(MOD(time, 12))
(1.00, 0.47), (2.00, 0.51), (3.00, 0.22), (4.00, 0.24), (5.00, 0.18), (6.00, -0.32), (7.00, -0.42), (8.00, -0.49), (9.00, -0.38), (10.0, -0.2), (11.0, 0.07), (12.0, 0.13)
Seas_norm_Latmax = GRAPH(MOD(time, 12))
(1.00, -1.00), (2.00, -1.00), (3.00, -1.00), (4.00, -1.00), (5.00, 2.17), (6.00, 2.51), (7.00, 0.63), (8.00, 0.24), (9.00, 0.05), (10.0, -0.03), (11.0, -0.66), (12.0, -0.92)
Seas_norm_Latmin = GRAPH(MOD(time, 12))
(1.00, 1.04), (2.00, 1.37), (3.00, 0.56), (4.00, 0.38), (5.00, -0.29), (6.00, -0.23), (7.00, -0.62), (8.00, -0.71), (9.00, -0.79), (10.0, -0.74), (11.0, -0.28), (12.0, 0.32)
Seas_norm_Qmax = GRAPH(MOD(time, 12))
(1.00, -0.71), (2.00, -0.48), (3.00, -0.17), (4.00, -0.17), (5.00, 0.62), (6.00, 1.74), (7.00, 0.52), (8.00, 0.09), (9.00, -0.16), (10.0, -0.2), (11.0, -0.63), (12.0, -0.44)
Seas_norm_Qmin = GRAPH(MOD(time, 12))
(1.00, 0.58), (2.00, 0.81), (3.00, 0.84), (4.00, 1.58), (5.00, -0.1), (6.00, -1.00), (7.00, -1.00), (8.00, -1.00), (9.00, -0.82), (10.0, -0.56), (11.0, 0.11), (12.0, 0.54)
Temp_norm = GRAPH(mod(time, 12))
(1.00, -8.00), (2.00, -2.00), (3.00, 0.00), (4.00, 2.00), (5.00, 8.00), (6.00, 20.0), (7.00, 8.00), (8.00, 2.00), (9.00, 0.00), (10.0, -2.00), (11.0, -8.00), (12.0, -20.0)

5.2. Equations for the Coastal Model

A compilation of all equations in the radiocesium model for coastal areas using data for the default area given in Table 2.4.

Compartment: A-sediments
Amount_in_A_sediments(t) = Amount_in_A_sediments(t - dt) + (Sed_on_A + Resusp_to_A - Diffusion - Transport_to_passive_sediments) * dt
INFLOWS:
Sed_on_A = Amount_in_water*Sed_rate*(1-ET_areas)
Resusp_to_A = Amount_in_ET_sediments*Resuspension_rate*(Form_factor/3)
OUTFLOWS:
Diffusion = Amount_in_A_sediments*Diffusion_rate
Transport_to_passive_sediments = Amount_in_A_sediments*(1/(Age_of_active_sed)+Physical_decay_rate_for_Cs)

Compartment: ET-sediments
Amount_in_ET_sediments(t) = Amount_in_ET_sediments(t - dt) + (Sed_on_ET - Resuspension_to_water - Resusp_to_A - Decay_from_ET) * dt
INFLOWS:
Sed_on_ET = Amount_in_water*Sed_rate*(ET_areas)
OUTFLOWS:

Resuspension_to_water = Amount_in_ET_sediments*Resuspension_rate*(1-Form_factor/3)
Resusp_to_A = Amount_in_ET_sediments*Resuspension_rate*(Form_factor/3)
Decay_from_ET = Amount_in_ET_sediments*Physical_decay_rate_for_Cs

Compartment: inflow areas
Amount_in_inflow_areas(t) = Amount_in_inflow_areas(t - dt) + (Deposition_on_IA - Secondary_input_from_IA - Decay_from_IA - Transport_from_IA_to_OA) * dt
INFLOWS:
Deposition_on_IA = Fallout_of_Cs*Catchment_area*(1-Outflow_areas_OA)
OUTFLOWS:
Secondary_input_from_IA = (Estuary*Amount_in_inflow_areas*Default_runoff_rate_for_OA*(1-Distribution_coefficient))/(Soil_permeability_factor*12)
Decay_from_IA = Amount_in_inflow_areas*Physical_decay_rate_for_Cs
Transport_from_IA_to_OA =
(Amount_in_inflow_areas*Default_runoff_rate_for_OA*Distribution_coefficient)/(12*Soil_permeability_factor)

Compartment: outflow areas
Amount_in_outflow_areas(t) = Amount_in_outflow_areas(t - dt) + (Deposition_on_OA + Transport_from_IA_to_OA - Secondary_input_from_OA - Decay_from_OA) * dt
INFLOWS:
Deposition_on_OA = Fallout_of_Cs*Catchment_area*Outflow_areas_OA
Transport_from_IA_to_OA =
(Amount_in_inflow_areas*Default_runoff_rate_for_OA*Distribution_coefficient)/(12*Soil_permeability_factor)
OUTFLOWS:
Secondary_input_from_OA = Estuary*Amount_in_outflow_areas*Modified_runoff_rate
Decay_from_OA = Amount_in_outflow_areas*Physical_decay_rate_for_Cs

Compartment: water
Amount_in_water(t) = Amount_in_water(t - dt) + (Inflow_from_sea + Direct_fallout + Diffusion + Resuspension_to_water + Secondary_input_from_IA + Secondary_input_from_OA - Outflow - Sed_on_A - Sed_on_ET) * dt
INFLOWS:
Inflow_from_sea = Conc_in_Sea*(Sur_water_flow+Deep_water_flow)
Direct_fallout = Fallout_of_Cs*Coastal_area
Diffusion = Amount_in_A_sediments*Diffusion_rate
Resuspension_to_water = Amount_in_ET_sediments*Resuspension_rate*(1-Form_factor/3)
Secondary_input_from_IA = (Estuary*Amount_in_inflow_areas*Default_runoff_rate_for_OA*(1-Distribution_coefficient))/(Soil_permeability_factor*12)
Secondary_input_from_OA = Estuary*Amount_in_outflow_areas*Modified_runoff_rate
OUTFLOWS:
Outflow = Amount_in_water*Outflow_rate
Sed_on_A = Amount_in_water*Sed_rate*(1-ET_areas)
Sed_on_ET = Amount_in_water*Sed_rate*(ET_areas)

Compartment: Cs-concentration in fish
Cs_concentration_in_fish(t) = Cs_concentration_in_fish(t - dt) + (Biouptake - Excretion) * dt
INFLOWS:
Biouptake =
Aut_prod_moderator*Salinity_moderator_for_biouptake*K_moderator*Biomagnification_factor*SMTH1((Dissolved_conc+Total_particulate_conc),(Biouptake_delay_factor))
OUTFLOWS:
Excretion = Cs_concentration_in_fish*(Biological_halflife+Physical_decay_rate_for_Cs)

A_areas = 1-ET_areas
Age_of_active_sed = Bioturbation_factor*Thickness_of_active_layer*(1-Water_content/100)*Bulk_density*10^6/(Gross_sed*30)
Aut_prod_moderator = if Inorganic_nitrogen_conc > 20 then 8 else (28-Inorganic_nitrogen_conc)
Averaging_function = (500000)/(Mean_annual_precipitation*(Latitude*Catchment_area)^(1/4)))

Biological_halflife = 30*(EXP(-6.583-0.111*LOGN(Fish_weight)+0.093*Surface_water_temp+0.326))
Biomagnification_factor = if Fish_weight > 10 then (if Fish_species_feed_habit = 0.005 then 250/Fish_weight else 0) + (if Fish_species_feed_habit = 0.5 then 2 else 0) + (if Fish_species_feed_habit = 0.05 then 3 else 0) + (if Fish_species_feed_habit = 0.02 then 2.5 else 0)+ (if Fish_species_feed_habit = 0.1 then 3.5 else 0) + (if Fish_species_feed_habit = 0.125 then 4 else 0)+(if Fish_species_feed_habit = 0.25 then 2.5 else 0)+ (if Fish_species_feed_habit = 0.125/2 then 3.5 else 0) + (if Fish_species_feed_habit = 0.01 then 1.5 else 0) + (if Fish_species_feed_habit = 0.2 then 2.5 else 0) else 25
Bioturbation_factor = if Gross_sed > 20 then 1 else (4-0.02*(Gross_sed/1-1))

Biouptake_delay_factor = if Fish_weight > 100 then ((Fish_weight)/100)^1 else 1
Bulk_density = 100*2.6/(100+(Water_content+Loss_on_ignition)*(2.6-1))

Catchment_area = 10*Coastal_area m^2
Coastal_area = 5*10^6 m^2
Conc_in_A_sed = Amount_in_A_sediments/((10^3)*Volume_of_A_sed*Bulk_density*(1-Water_content/100))
Conc_in_Sea = Initial_Cs_conc_in_sea*EXP(-TIME*Halflife_in_sea)
Cswa_Bqm^3 = 1000*Total_conc_in_water
Current_velocity = 2.5 cm/sec

Dres = (Resuspension_to_water+Resusp_to_A)/(Sed_on_A+Sed_on_ET)
Dcrit = if Dcrit_limit > Mean_depth then Mean_depth else Dcrit_limit
Dcrit_limit = if ABS(Surface_water_temp-Bottom_water_temp) < 4 then 0 else 1*ABS(Surface_water_temp-Bottom_water_temp)
Deep_water_flow = (Vold)/(Td+Mixing_rate)
Default_Kd-constant = 800000
Default_runoff_rate_for_OA = 0.04*(Mean_annual_precipitation/650)^2
Default_settling_velocity = if DC_resusp > 0.99 then 2*12/12 else ((12/12)*(1-DC_resusp)+2*(12/12)*(DC_resusp))
Diffusion_rate = (1*0.002/12)*(Gross_sed/10)*(0.02/Thickness_of_active_layer)
Dissolved_conc = Dissolved_fraction*Total_conc_in_water
Dissolved_fraction = 1/(1+(Kd*Susp_concentration)/1000000)
Distribution_coefficient = 0.5

Estuary = 1
ET_areas = if ETlimit1 < 0.15 then 0.15 else ETlimit1
ETlimit1 = If 10^(0.14-2.86*((Volume*10^(-9))^0.5)+0.08*log10(Mean_filter_factor)+0.02*Max_depth-0.25*Relative_depth) > 1 then 1 else 10^(0.14-2.86*((Volume*10^(-9))^0.5)+0.08*log10(Mean_filter_factor)+0.02*Max_depth-0.25*Relative_depth)
Exposure = 100*Section_area/Coastal_area
Filter_factor = 3
Fish_species_feed_habit = if Fish_weight > 10 then 1/8 else 0.005
Fish_weight = 1000 g ww

Form_factor = 3*Mean_depth/Max_depth
Gross_sed = if ((Form_factor/3)*10^(0.323-0.611*LOG10(Coastal_area*10^(-6))+1.009*LOG10(Inorganic_nitrogen_conc)-0.587*LOG10(Secchi_depth)+0.0238*Filter_factor)) < 1 then 1 else ((Form_factor/3)*10^(0.323-0.611*LOG10(Coastal_area*10^(-6))+1.009*LOG10(Inorganic_nitrogen_conc)-0.587*LOG10(Secchi_depth)+0.0238*Filter_factor))

Halflife_in_sea = 1/(3*12)

Initial_Cs_conc_in_sea = Fallout/(100)
Inorganic_nitrogen_conc = 20

K_concentration = 1 mg/l
K_moderator = 300/(4.5+(K_concentration/0.0391)^0.75)
Kd = Default_Kd/K_concentration

Latitude = 60 °N
Loss_on_ignition = 10*(100-Water_content)/100 % dw

Max_depth = 30 m
Mean_annual_precipitation = 750 mm/yr
Mean_annual_Q = 0.38*Mean_annual_precipitation*Catchment_area*10^-6/3600/24/365.25*1000
Mean_depth = 9 m
Mean_filter_factor = Filter_factor/Number_of_openings
Mean_slope = 4.59 + 0.19*Mean_depth - 1.86*((Section_area*1000)/Volume)^0.5 + 5.7*LOG10(Relative_depth)
Mixing_rate = if 1/ABS(Surface_water_temp-Bottom_water_temp) > 1 then 1 else 1/ABS(Surface_water_temp-Bottom_water_temp)
Modified_runoff_rate = Seasonal_moderator_for_Q*Default_runoff_rate_for_OA/(12*(sqrt(DELAY(TIME,Month_of_fallout,1))))
Month_of_fallout = 5
Monthly_discharge_Q = Mean_annual_Q*60*60*24*30*Seasonal_moderator_for_Q

Number_of_openings = 2

Outflow_areas_OA = 0.2
Outflow_rate = ((Total_water_exchange/Volume)+Physical_decay_rate_for_Cs)

Particulate_amount = Amount_in_water*(1-Dissolved_fraction)
Physical_decay_rate_for_Cs = 0.693/(30.2*12)
Predicted_Td_in_days = if EXP(7.72-2.93*(Mean_filter_factor)-19.16/Mean_slope-0.6*LOGN(Exposure)) < 1 then 1 else EXP(7.72-2.93*(Mean_filter_factor)-19.16/Mean_slope-0.6*LOGN(Exposure))
Predicted_Ts_rate = if 1*(1/Ts_bay)+1*(1/Ts_from_tide)+1*(1/Ts_from_coastal_current) > 30 then 30 else 1*(1/Ts_bay)+1*(1/Ts_from_tide)+1*(1/Ts_from_coastal_current)

Relative_depth = Max_depth*(3.1416^0.5)/(20*((Coastal_area*10^(-6))^0.5))
Resuspension_rate = 30/21
RR_deep = Deep_water_flow/(Vold+1)

Sal_moderator_for_sed = (1+0.1*(Salinity/1-1))
Salinity = 6 ‰
Salinity_moderator_for_biouptake = if (0.9-0.1*((Salinity/1)-1)) < 0.2 then 0.2 else (0.9-0.1*((Salinity/1)-1))
Seasonal_moderator_for_Q = 1+0.526*((Latitude-35)^2.18/35^2.18*Seas_norm_Latmax+(1-(Latitude-35)^2.18/35^2.18)*Seas_norm_Latmin)+
0.265*(Mean_annual_Q^0.22/5000^0.22*Seas_norm_Qmin+(1-Mean_annual_Q^0.22/5000^0.22)*Seas_norm_Qmax)
Secchi_depth = 5 m
Section_area = 10*1000 m^2
Sed__velocity = Default_settling_velocity*Susp_moderator*Sal_moderator_for_sed
Sed_rate_for_part_Cs2 = Sed_rate_for_part_Cs1*(1-Dissolved_fraction)
Sed_rate_of_part_Cs1 = Sed__velocity/Mean_depth
Soil_permeability_factor = 25
Sur_water_flow = (Vols*Predicted_Ts_rate)
Susp_concentration = 3 mg/l
Susp_moderator = 1+0.22*(Susp_concentration/3.3-1)

Td = if Td_in_months < 3/30 then 3/30 else 1*Td_in_months+0*10/Predicted_Ts_rate
Td_in_months = if Predicted_Td_in_days/30 > 4 then 4 else Predicted_Td_in_days/30
Thickness_of_active_layer = 0.05 m
Tidal_amp = 2 cm/day
Total_conc_in_water = 0.001*Amount_in_water/Volume
Total_particulate_conc =
0.001*(Fish_species_feed_habit*(Amount_in_A_sediments+Amount_in_ET_sediments+Particulate_amount))/Volume
Total_water_exchange = Monthly_discharge_Q+Sur_water_flow+Deep_water_flow
Ts_bay = if Ts_for_coastal_bay ≥ Ts_from_coastal_current then 100 else Ts_for_coastal_bay
Ts_for_coastal_bay = if Exposure ≥ 1.3 then Ts_from_coastal_current else Ts_limit_1
Ts_from_coastal_current = if Exposure < 1.3 then 100 else ((Vols)/(Ycurrent*Current_velocity*0.01*60*60*24*30*0.5*Section_area))
Ts_from_tide = (Vols)/(Coastal_area*Tidal_amp*Ytide*0.01*30)
Ts_limit_1 = if (EXP(3.49-4.33*(Exposure^0.5)))/(30) < 1/30 then 1/30 else (EXP(3.49-4.33*(Exposure^0.5)))/(30)

Vold = (Volume-Vols+1) m^3
Vols = if Dcrit = 0 then WB_area*Mean_depth else (WB_area*Dcrit+(Coastal_area-WB_area)*Dcrit/3))
Volume = Mean_depth*Coastal_area m^3
Volume_of_A_sed = Coastal_area*(A_areas)*Thickness_of_active_layer

Water_content = 75 % ww
WB_area = Coastal_area*((Max_depth-Dcrit)/(Max_depth+Dcrit*EXP(3-Form_factor^1.5)))^(0.5/Form_factor)

Ycurrent = if YEx*YMFf > 1 then 1 else YEx*YMFf
YEx = (1+0.5*((Exposure/10)-1))
YMFf = (1+0.5*((Mean_filter_factor/50)-1))
Ytide = if YEx*YMFf > 1 then 1 else YEx*YMFf

Bottom_water_temp = GRAPH(MOD(TIME,12))
(1.00, 4.00), (2.00, 4.00), (3.00, 4.00), (4.00, 4.00), (5.00, 4.50), (6.00, 6.00), (7.00, 8.00), (8.00, 10.0), (9.00, 8.00), (10.0, 7.00), (11.0, 4.00), (12.0, 4.00)
Fallout_of_Cs = GRAPH(TIME)
50,000 Bw month 5
Seas_norm_Latmax = GRAPH(MOD(time, 12))

(1.00, -1.00), (2.00, -1.00), (3.00, -1.00), (4.00, -1.00), (5.00, 2.17), (6.00, 2.51), (7.00, 0.63), (8.00, 0.24), (9.00, 0.05), (10.0, -0.03), (11.0, -0.66), (12.0, -0.92)
Seas_norm_Latmin = GRAPH(MOD(time, 12))
(1.00, 1.04), (2.00, 1.37), (3.00, 0.56), (4.00, 0.38), (5.00, -0.29), (6.00, -0.23), (7.00, -0.62), (8.00, -0.71), (9.00, -0.79), (10.0, -0.74), (11.0, -0.28), (12.0, 0.32)
Seas_norm_Qmax = GRAPH(MOD(time, 12))
(1.00, -0.71), (2.00, -0.48), (3.00, -0.17), (4.00, -0.17), (5.00, 0.62), (6.00, 1.74), (7.00, 0.52), (8.00, 0.09), (9.00, -0.16), (10.0, -0.2), (11.0, -0.63), (12.0, -0.44)
Seas_norm_Qmin = GRAPH(MOD(time, 12))
(1.00, 0.58), (2.00, 0.81), (3.00, 0.84), (4.00, 1.58), (5.00, -0.1), (6.00, -1.00), (7.00, -1.00), (8.00, -1.00), (9.00, -0.82), (10.0, -0.56), (11.0, 0.11), (12.0, 0.54)
Surface_water_temp = GRAPH(MOD(TIME,12))
(1.00, 0.00), (2.00, 0.00), (3.00, 3.00), (4.00, 4.00), (5.00, 7.00), (6.00, 11.0), (7.00, 16.0), (8.00, 20.0), (9.00, 15.0), (10.0, 7.00), (11.0, 4.00), (12.0, 1.00)

INDEX

A

B

C

D

E

F

G

K

L

M

N

O

P

V

X

Y

Z

Å

Ø

GPSR Compliance

The European Union's (EU) General Product Safety Regulation (GPSR) is a set of rules that requires consumer products to be safe and our obligations to ensure this.

If you have any concerns about our products, you can contact us on ProductSafety@springernature.com

In case Publisher is established outside the EU, the EU authorized representative is:

Springer Nature Customer Service Center GmbH
Europaplatz 3
69115 Heidelberg, Germany

Batch number: 10607222

Printed by Printforce, the Netherlands